TREE OF ORIGIN

TREE OF ORIGIN

What Primate Behavior
Can Tell Us about
Human Social Evolution

Frans B. M. de Waal, Editor

HARVARD UNIVERSITY PRESS

Cambridge, Massachusetts, and London, England

Unless otherwise noted, illustrations by Margaret C. Nelson

Second printing, 2002

First Harvard University Press paperback edition, 2002

Library of Congress Cataloging-in-Publication Data

Tree of origin : what primate behavior can tell us about human social evolution /
Frans B.M. de Waal, editor.
    p. cm.
Includes bibliographical references and index.
ISBN 0-674-00460-4 (cloth)
ISBN 0-674-01004-3 (pbk.)
1. Primates—Behavior. 2. Human evolution. 3. Social evolution.
I. Waal, F. B. M. de (Frans B. M.), 1948–

QL737.P9 T75 2001
599.8'15—dc21   00-050581

Contents

TREE OF ORIGIN

Frans B. M. de Waal

Introduction

I︎T IS READILY APPARENT that comparisons between humans and other animals take two basic forms, depending on whether the main objective is to confirm human identity or to stress the common thread that runs through all forms of life. The first approach is still commonplace in the social sciences and the humanities, where the importance of any particular subject is underlined by claims that ours is the only species to which the topic could possibly apply. The second approach—at least as old— received strong impetus from Darwin's theory of evolution by natural selection. Especially in the last couple of decades, the view of human behavior as the product of evolution, hence subject to the same explanatory framework as animal behavior, has gained ground and respectability. It has been transformed from a controversial minority view to one that today is widely applied.

This transformation was made possible by major theoretical developments in the field of animal behavior—such as the explanations of cooperative behavior at the heart of sociobiology and behavioral ecology—and by the courage of a few scientists who in the 1970s began sending wake-up calls to those sections of the academic community traditionally favoring nurture over nature.

The evidence for a connection between genes and behavior is mounting rapidly. Twins-reared-apart studies have reached the status of common knowledge, and almost every week newspapers report a new human gene—those involved in schizophrenia, epilepsy, and Alzheimer's disease, even in behavioral traits such as thrill-seeking. We are also learning more about genetic and neurological differences between men and women. Not that there is a simple one-on-one relation between genes and behavior: our sound-bite culture tends to turn findings into a gene-for-this and a gene-for-that language, whereas in reality behavioral variability is only partly explained by genetic factors (the other part being attributable to the envi-

1

ronment). Nevertheless, the list of scientific advances implying genetic influences is getting longer by the day, resulting in a critical mass of evidence that is impossible to ignore. Understandably, academics who have spent a lifetime condemning the idea that biology influences human behavior are reluctant to change course, but they are being overtaken by the general public, which seems to have accepted that genes are involved in just about everything we do and are.

Another development that has influenced the way we look at our place in nature, and that has particular relevance to the present volume, is the proliferation of research on monkeys and apes. In terms of genetics, neurology, cognition, and social behavior, many cherished assumptions in support of human/animal dualism have fallen by the wayside. DNA studies have placed the apes far closer to us than anyone deemed possible at the beginning of the twentieth century. Not only are chimpanzees and bonobos genetically our closest relatives, the reverse is also true; that is, chimpanzees and bonobos are closer to us than to, say, gorillas.

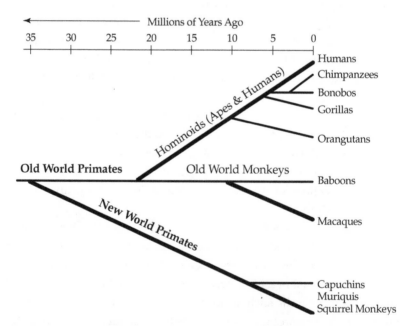

Figure I.1 Evolutionary tree showing the main branches of the primate order (except for the prosimians): the New World monkeys, the Old World monkeys, and the hominoid lineage that produced our own species. The diagram reflects advances in DNA analysis that place the African apes (gorillas, bonobos, and chimpanzees) much closer to humans than previously suspected.

People have a tendency to lump all two hundred primate species under the rubric of "monkeys," but the great apes (only four species: chimpanzee, bonobo, gorilla, and orangutan) together with gibbons and humans are a distinct group, the Hominoids, not to be confused with the Old World monkeys (such as baboons and macaques), which in turn are not to be confused with New World monkeys (such as capuchins and marmosets). Figure I.1 and Table I.1 offer help with the classification. Readers are referred to MacDonald (1984) and Napier and Napier (1985) for more detailed primate taxonomies.

So, after a period in which human behavior has been freely interpreted in light of what we know about pigeons (such as the research of the learning psychologist B. F. Skinner), geese (Konrad Lorenz), social insects (E. O. Wilson), and rodents (Robert Ardrey, who compared people in inner cities with crowded rats), comparisons increasingly focus on species that are a lot closer and more similar to us than any of the above.

Such comparisons have captured public attention ever since Desmond

Table I.1 The hominoid family is characterized by flat chests, rotational ability in the shoulders, absence of tails, and relatively large body size. It includes the Asian gibbons and siamangs (the so-called lesser apes), the Pongidae (great apes) of Africa and Asia, and modern humans. With the exception of the gibbon family, this table delineates the species and subspecies of extant humans and apes, customarily divided into five different species belonging to four genera.

Genus	Species	Subspecies	Common name
Gorilla	gorilla	beringei	Mountain gorilla
		graueri	Eastern lowland gorilla
		gorilla	Western lowland gorilla
Pongo	pygmaeus	pygmaeus	Bornean orangutan
		abelii	Sumatran orangutan
Pan	troglodytes	verus	Masked or pale-faced chimpanzee (west Africa)
		troglodytes	Black-faced chimpanzee (central Africa)
		schweinfurthii	Long-haired chimpanzee (eastern Africa)
	paniscus		Bonobo (previously: pygmy chimpanzee)
Homo	sapiens	sapiens	Modern human

Morris in 1967 published *The Naked Ape*, followed in 1971 by Jane Goodall's *In the Shadow of Man*. Both books relied on direct ape-human comparisons—the former from the human perspective; the second, more subtly, via an in-depth account of chimpanzee social life with the implicit message that these animals are a great deal like us. These two early books are only the tip of the iceberg, however. A steady stream of scholarly and popular publications have appeared since, detailing numerous parallels between human and ape behavior. Often the implication has been that these parallels represent homologies; that is, they derive from the common ancestor of humans and apes. Some of the contributors to the present volume have been in the forefront of this literature.

The growing influence of primate research has forced linguists to look more closely at their definitions of language, and anthropologists to consider nonhuman culture. Ape studies have also triggered the current fascination with self-awareness and so-called Theory of Mind in child psychology. Even the origins of human politics, warfare, and morality are now being discussed in the light of primate observations.

What is the objective of such comparisons? Is it just to show that we are not alone in being cognitively and socially complex, or does it go deeper? If humans and other primates show similar behavior, does that make the apes look smarter, assuming that they put as much cognition into their actions as we do? Or does it make us look dumber? Do similarities prove that we are instinctive creatures held on a leash by our biology? These are the sorts of questions that fueled many debates in the 1970s. There were accusations directed at biologists of genetic determinism and questions about their application of human concepts to animal behavior (for instance, Sherwood Washburn's 1978 essay "What We Can't Learn about People from Apes"). The current generation of primatologists, however, is quite averse to attempts to squeeze behavior into a simple "learned versus innate" dichotomy. We assume that *all* behavior in *all* primates, including our own species, derives from a combination of evolved tendencies, environmental modification, development, learning, and cognition. The word *instinct*, with its connotation of a purely genetic derivation, has indeed become outdated except in usage such as a "language instinct," where the concept has been broadened to stand for learning predisposition.

In sum, this book is not about our species as a preprogrammed robot destined by its biology to act one way or another. The contributors have far too high an opinion of the behavioral flexibility of their primary research subjects, which in all cases are nonhuman primates, to embrace such a view. If they see monkeys and apes as mentally complex, socially skilled, even cultural beings, why would they want to simplify human behavior? Rather, the goal is to understand human social evolution from the perspective of what we know about the social organization, communication, learned habits, subsistence, reproduction, and cognition of other extant primates. This is no simple task, and it certainly has no simple answers.

The genesis of our book goes back to October 1997, when a conference on Human Evolution was held at the well-known Cold Spring Harbor Laboratory in New York State. A well-attended and well-received session, organized by the writer, was entitled "Primate Behavior and the Reconstruction of Human Social Evolution." Six of the nine contributors to this volume participated. All the other sessions, except for one on paleontology, dealt with molecular genetics.

It was felt that a symposium topic such as primate behavior deserved its own volume; not since Warren Kinzey's 1987 book, *The Evolution of Human Behavior: Primate Models,* has anyone brought together primatologists specifically around the question of human evolution. Most primatologists stay close to their data without venturing too far into speculation about the origins of human behavior. Even if some of our contributors are known exceptions, there are no works juxtaposing their views with those of others. In putting this volume together, we followed three simple rules:

1. All authors are behavioral primatologists—psychologists, anthropologists, or zoologists extensively familiar with the behavior of monkeys and apes. Informed opinions about human evolution are to be found in many quarters (neuroscience, paleontology, linguistics), but we wished to present the primate perspective straight from the pen of people with intimate, firsthand knowledge of the natural or naturalistic behavior of this special taxonomic group.

2. All contributors speculate about the origins of human evolution. How is our species similar to or different from other primates, and which selection pressures may have forced our ancestors to evolve in the direction they took. Some authors address this issue more explicitly than others, but all present an aspect of what primate behavior can tell us about human social evolution.

3. Contributors were asked to write in an accessible, jargon-free style. Only the notes provide technical background. The authors should be for-

given for not citing each and every relevant study, and for not mentioning every angle of a problem. Even though all contributors are recognized experts in their field, there is no mainstream view on the topic of human evolution. Hence this book should be supplemented with other readings, including alternative views by other scholars.

No attempt has been made here to reconcile differences of opinion. It would be surprising if any treatment of this topic at this stage could provide lasting answers. Our hope, rather, is that this book will make readers rethink human evolution, and consider peculiarities of our species or lines of argument that they have not taken into account before. Even if what one author considers obvious is questioned by another, it is the underlying *problem* that matters most.

For example, Robin Dunbar follows many other primatologists in arguing for social complexity as the driving force behind the evolution of intelligence, whereas Richard Byrne thinks that, at least in the apes, planning of action chains played a significant role—and Craig Stanford sees the politics surrounding meat-sharing as yet another factor. All three authors try to come up with reasons why primate brains expanded so much more, relative to body size, than the brains of other land mammals. The reader need not choose among these points of view to recognize that an evolutionary explanation is needed, which at the same time may tell us something about our own species, which has the largest brain of all.

Other differences of opinion are evident too. Richard Wrangham regards the chimpanzee as the best model of the last common ancestor of humans and apes, whereas Frans de Waal argues that the bonobo is at least as good a candidate. Stanford emphasizes hunting and meat-sharing in human evolution, in contrast to Wrangham's focus on plant foods. Charles Snowdon looks for continuities between primate communication and human language, whereas Dunbar emphasizes the social bonding function of language and places it functionally closer to grooming. Finally, whereas many authors emphasize the behavior and cognitive capacities of our closest relatives, Karen Strier urges us to look beyond the apes.

This final point relates to two different approaches to the matter of human evolution. One involves development of general principles that apply to a great variety of species, from insects to birds to humans. Primates are a not particularly numerous or varied group of organisms to consider when we investigate how the distribution of food affects social organization, or the conditions under which animals increase in body size, or how competitive tendencies are balanced with the need for cooperation. Kin selection is an example of an explanatory principle that has helped us understand a multitude of social phenomena in a wide range of animals.

The other approach is to trace phylogeny. For example, with regard to human tool use, we may ask whether our ancestors were likely to have similar capacities; we find part of the answer in the fact that the great apes are proficient tool users. This makes it likely that our direct ancestors used tools as well. In this particular approach—the historical reconstruction of human evolution—apes are an absolutely critical link. Even so, to realize how special the great apes are we still need to consider them in the larger context of the primate order and the animal kingdom as a whole.

We follow both approaches here, sometimes appealing to biology in general, at other times to direct comparisons with our closest relatives. The result is a book that revolves around the following themes:

Ecology. Natural ecology, food distribution, hunting, and the role of subsistence in shaping social organization and cognition are discussed by Anne Pusey, Craig Stanford, Richard Wrangham, and Richard Byrne. William McGrew reviews material culture, which is an integral part of food acquisition and processing.

Sex and Reproduction. Reproduction, inbreeding avoidance, and genetics are discussed by Anne Pusey and Karen Strier, whereas Frans de Waal discusses the nonreproductive functions of sexual behavior.

Social Organization. Territoriality between groups, the organization within groups, and the need for conflict resolution and reconciliation are reviewed by Anne Pusey, Frans de Waal, and Karen Strier.

Social Sophistication and Cognition. Group organization, its cognitive implications, and the evolution of a large neocortex are treated by Craig Stanford, Richard Byrne, and Robin Dunbar. To this Charles Snowdon adds a review of how the natural communication systems of primates relate to human language.

Hominization. The process of hominization that turned our ancestors from apelike primates into humanlike ones, is most explicitly discussed by Richard Wrangham, Robin Dunbar, and William McGrew.

It is around these five themes that we weave arguments about how our species fits, or does not fit, with current knowledge about other primates. The resulting picture is neither complete nor definitive, but it will no doubt stimulate discussion. Our hope is that it may inspire others to look more closely at both primate behavior and the human fossil record in order to adjust old scenarios of human evolution or develop new ones.

1

Of Genes and Apes: Chimpanzee Social Organization and Reproduction

Anne E. Pusey

For many decades, chimpanzees have been observed at Gombe National Park and other African field sites. Now DNA analyses have begun to further enrich our understanding of their social life. Chimpanzee behavior and ecology are reviewed here, along with the genetic implications for kinship, paternity, and inbreeding.

WHAT KIND OF SOCIAL groups did our ancestors live in? If we had an idea of the social structure of our apelike ancestors, we would have a starting point from which to trace the evolutionary changes that have led to the unique characteristics of humans. Many anthropologists believe that chimpanzees may provide an appropriate model for such an ancestor. In this chapter I review our current knowledge of chimpanzee social structure as it has been revealed by long-term studies of chimpanzee communities and by field studies of the genetic relationships of individuals within populations.

Chimpanzees are our closest living relatives, sharing close to 99 percent of our DNA. They are more closely related to us than they are to gorillas. By analogy with the taxonomy of other closely related species, the biologist Jared Diamond argues that we should be placed in the same genus as chimpanzees and bonobos, making us the "third chimpanzee." Despite obvious physical differences, many aspects of our anatomy and physiology are almost identical because of this genetic similarity. Chimpanzees also resemble us in their longevity, a fact that has long made it difficult to understand key features of their social structure. Now, however, several studies have sufficient duration to reveal chimpanzee patterns of residence, kin relationships, and mating patterns. These features of social structure influence patterns of cooperation and social behavior. Also, advances in genetic techniques have made it possible to measure kin relations from DNA directly. While the risk of casualty or death during capture for the collection of genetic material has been considered unacceptable in endangered species such as chimpanzees, new techniques have made it possible to extract DNA from noninvasively collected samples, such as hair shed in night nests, feces, and pellets of chewed, discarded food. Preliminary results have already yielded some surprising insights.

History of Chimpanzee Studies

The anatomic similarity of chimpanzees to humans has been recognized from antiquity, and observations of their behavior in captivity, especially in the early part of this century, revealed their intelligence. Despite various efforts beginning in the 1930s, very little was discovered for decades about their behavior in the wild, the sort of groups they lived in, or their way of life. Chimpanzees live in the remote forests of sub-Saharan Africa, and even when researchers penetrated those forests, they found the chimpanzees extremely wary of humans. Several short-term studies and surveys in various parts of Africa in the early 1960s provided some information on grouping and habits. But the first breakthroughs in understanding social structure came when biologists made concerted efforts to watch chimpanzees in the eastern part of their range in Tanzania, where the forest is less thick and is broken by grassland. In 1960 Louis Leakey sent Jane Goodall to observe chimpanzees in what is now Gombe National Park on the northeastern shore of Lake Tanganyika. Gradually, as the apes allowed her to approach more closely, she began to recognize and name individuals. When she started to provision the chimpanzees regularly with bananas, more and more of them came to the provisioning area until she was able to recognize and observe all the chimpanzees in the local population. At about the same time, Japanese scientists made several expeditions in areas to the south of Gombe, and in 1965 Toshisada Nishida established a field station in the Mahale Mountains. He provisioned the chimpanzees with sugarcane and also identified individuals. Both studies continue to this day and have contributed much of what we know of chimpanzee behavior.

Provisioning allowed the researchers to make close observations and habituate even the shyest chimpanzees, so that it was eventually possible to follow individuals wherever they went. The technique has been criticized, however, because of its possible effects on behavior. Long-term studies without provisioning have been established in other areas of Africa (see Figure 1.1) by Yukimaru Sugiyama at Bossou, Guinea, in 1976; by Christophe and Hedwige Boesch in the Taï Forest, Côte d'Ivoire, in 1979; and by Richard Wrangham in Kibale Forest, Uganda, in 1987. These have succeeded in habituating individuals and are producing data on social behavior comparable to those from provisioned sites. On the whole, the results are very similar.

Social Grouping Patterns

On entering a forest where chimpanzees live, you will immediately notice loud pant-hoots and perhaps drumming on trees. If you follow these sounds, you may encounter a group of twenty or more chimpanzees on the ground or in the trees—including females with melon-sized pink genital swellings, and males gathering round them and mating in turn. Excitement and aggression among the males will include charging displays in which they drag branches and throw rocks. You will also see females with nursing infants on their bellies, and juveniles and adolescents engaged in play sessions with raucous laughter. But if you stay in the forest for many days, you will encounter much smaller groups, and even lone individuals feeding quietly in the forest.

By recognizing individuals and watching them over many years, we have gradually come to understand the nature of their complex society. In the early years of her study, Goodall observed that individual chimpanzees moved alone or in small temporary parties of any combination of age and sex, and only sometimes gathered in larger groups. She believed that the only stable association between individuals was that between a mother

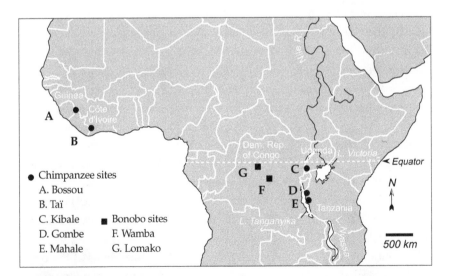

Figure 1.1 Equatorial Africa, showing sites of long-term field research on chimpanzees (Mahale Mountains and Gombe, both Tanzania; Kibale, Uganda; Taï, Côte d'Ivoire; Bossou, Guinea) and bonobos (Wamba and Lomako, both Democratic Republic of Congo).

and her offspring. However, after just three years at Mahale, Nishida concluded that although they were almost never seen all together, chimpanzees lived in larger "unit groups" (later called communities by Western scientists), consisting of several adult males and a similar or larger number of females and their offspring. He observed an annual seasonal displacement of one community, the K-group, by the larger M-group, and surmised that relations between them were antagonistic. We now know that wherever they are observed, chimpanzees live in communities consisting of anywhere from 20 to 120 individuals.

Within the community the sexes differ in their sociability and ranging patterns, and the behavior of females depends on their reproductive state. At intermittent periods during their lives, females undergo cycles of sexual receptivity advertised by sexual swellings that last about 13 days per 36-day cycle. They cycle for about two years at adolescence until they conceive, and then for a few months before the conception of each successive infant. Most females cycle once or twice after conception. The average interbirth interval between offspring that survive to weaning is 5.2 years at Gombe (7 years at Kibale), and during most of this time females are not receptive.

When females have swellings, they are extremely gregarious and range over large areas. Otherwise females usually feed alone or are accompanied only by their dependent offspring in small core areas that overlap but are distinct from those of other females (Figure 1.2). Gombe females spend about 50 percent of their time with only their dependent offspring in core areas of about 2 square kilometers. In contrast, the adult males are more social, spending only 18 percent of their time alone, and they typically range over an area of 8–15 square kilometers annually. Although females tend to be more localized, most do visit all areas of the community range at some point in their lives, either when they are sexually receptive or when they are drawn to fruiting trees outside their normal range. Similar sex differences in ranging and association are seen at Mahale and Kibale. However, at times the Taï Forest females appear to spend less time alone and range more widely with the males.

This kind of fission-fusion society, in which the individuals of the community spend some time alone and frequently join and leave temporary subgroups, is unusual among primates and mammals in general. It does occur in lions, hyenas, elephants, spider monkeys, and humans. Starting with the Japanese researchers, all scientists have identified food availability and cycling females as factors significantly influencing party size. Rich-

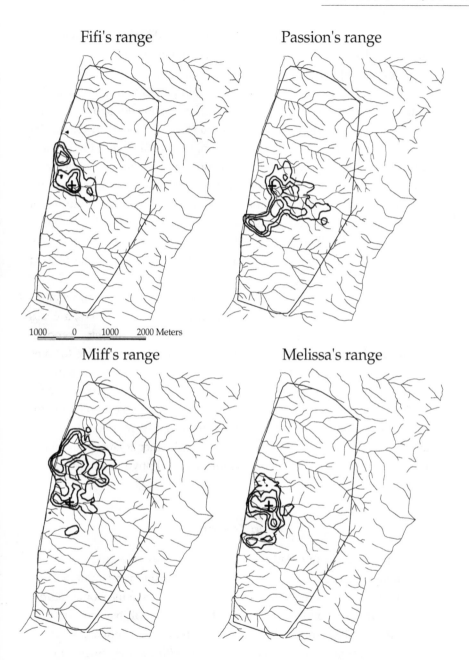

Figure 1.2 Ranges of the most frequently observed female chimpanzees at Gombe relative to the community range, 1975–1992. Female ranges consist of contours surrounding 90 percent, 80 percent, and 50 percent of the points recorded during follows of females while traveling with only their dependent offspring. The cross indicates the location of the provisioning station. Passion's larger range includes a southern portion that she abandoned in 1979 following a severe attack on her family, probably by chimpanzees of a neighboring community.

ard Wrangham, now an anthropologist at Harvard University, previously studied the feeding ecology of chimpanzees at Gombe. He has developed hypotheses to explain the unusual ranging patterns of chimpanzees on the basis of diet and reproductive strategy. Chimpanzees primarily eat ripe fruit, as well as smaller quantities of young leaves, flowers, shoots, pith, insects, and meat. The chimpanzees at Gombe feed on more than 120 species of plants, whose fruiting patterns are highly seasonal. Trees of many species only contain a few ripe fruits on any one day; therefore, the more individuals that feed together, the more quickly they have to leave one tree and seek others. Noting that mothers travel slowly because of the burden of their infant, Wrangham suggests that they minimize the feeding competition and maximize their feeding efficiency, and hence gain reproductive success by feeding alone in an area large enough to provide sufficient food sources.

Faced with this distribution of females in distinct core areas, what do males do? Because most male mammals, including chimpanzees, are less involved with the rearing of their infants than are females, their reproductive success in general depends more on their access to receptive females than on their feeding efficiency. By ranging over wider areas than females, males increase the number of potentially receptive females they can contact. Yet, because the distances are so great, single males cannot maintain exclusive access to a number of females. Instead, several males travel together over a large area and cooperate with each other against other males to protect their access to the larger number of receptive females they engage.

Group Territoriality

Although Nishida suggested as early as 1968 that intergroup relations were antagonistic, reliable observations of intercommunity interactions were not made until some time later. In the early years of the Gombe study, observations were concentrated at the feeding area of just one community. In the late 1960s researchers started to follow the chimpanzees more widely, and occasionally they encountered individuals from other communities. But these fled from the humans, and few intercommunity interactions were observed. There was no evidence that chimpanzees showed lethal aggression toward other members of their species.

This view changed in the early 1970s, with a series of pivotal observations at Gombe described by David Bygott, Goodall, Wrangham, and others. By 1970 the habituated community that visited the feeding station had begun to split into two subgroups. Over the next few years the divisions became more extreme. The subgroups were ultimately viewed as separate communities: the Kasekela in the north and the Kahama in the south. The Kahama community ceased to visit the feeding area, and the Kasekela community rarely traveled far into the southern part of the community range. When the males approached the boundary between the two ranges, they became cautious and silent, often pausing to listen and gaze into the range of their neighbors or to sniff the vegetation. This behavior is known as patrolling. When parties from the two communities heard each other or met, intense excitement and aggressive displays took place, with the larger of the two groups generally holding its ground or both groups withdrawing. Sometimes when a party heard calls from the other community, it would approach stealthily and silently. On one such occasion the Kasekela community came upon a single Kahama male with a female.

What ensued permanently changed our view of chimpanzees. The Kasekela males mounted a vicious gang attack on the male. As one male held the victim down, pinning him to the ground, others hit and bit him. The brutal attack lasted for ten minutes, and the male was mortally wounded. Over the next two years, several more lethal gang attacks were observed on single members of the Kahama community, some deep within that community's range, until none were left. The result was that the Kasekela community expanded its range into the south. Soon, however, it was pushed back by a still stronger community from the south, and the community from the north also started encroaching on its territory. Over the same period each community engaged in patrols and aggressive intergroup interactions with neighboring unhabituated communities.

Aggressive intercommunity interactions have since been observed in all the other long-term studies. At Mahale all the males of the small K-group disappeared; it was suspected that males of the larger neighboring M-group killed them, although no lethal attacks were observed.

It is clear from these observations that male chimpanzees cooperate to participate in intergroup hostility, but views of exactly what the males are defending or gaining by this behavior have changed over the years. Are males defending the integrity of the community's feeding territory and protecting their females from other males, or are they striving to gain access to more females by expanding their territory? Implicit in the original ideas of Nishida and other Japanese researchers was that both males and females belonged to a particular community (the bisexual community hypothesis). In 1979 Wrangham proposed an alternative view, the male-only community hypothesis. Noting that some females appeared to straddle the ranges of two communities, he suggested that instead of belonging to a particular community, females were distributed continuously over the habitat in dispersed core areas that depended simply on the position of other females. The male group then attempted to defend an area that encompassed as many females as possible. If the males could expand their territory, they would overlap the core areas of and gain access to more females. So, as the community range expanded, the number of adult females associating with the males would immediately increase.

Since Wrangham formulated this idea, evidence against it has accumulated. First, in addition to showing aggression to strange males, when males at the edges of the territory encounter stranger females who are not sexually receptive, they often attack them severely—even lethally. Goodall described severe attacks between 1971 and 1982 on 20 different stranger females at Gombe, 15 of whom had infants or juveniles. In three attacks the males killed the females' infants, but in most cases the aggression was directed primarily at the female. In 13 of 14 cases in which the female was well observed, she incurred serious wounds or left a great deal of blood. In no case were the males observed associating with these females after the attack. In addition, the Kasekela males killed one old Kahama female, then led observers to the wounded, dead body of another female.

Second, females are sensitive to the position of the community border. At Kibale, Wrangham and his colleagues Colin Chapman and Isabyre Basuto found that females avoid traveling in the periphery of the community range. At Gombe, Jennifer Williams and I have found that the position of female core areas is generally very stable. However, the size of the core areas of most females expands and contracts as the community range expands and contracts, and some females have abandoned portions of their core areas when the community boundary shifted. Goodall describes how one female, Passion, permanently abandoned the southern portion of her range after she and her family had been severely attacked, probably by members of the adjacent community. Finally, at Mahale, following the disappearance of their community males, most K-group females changed their ranges to associate with the M-group, again showing that female ranging patterns are often sensitive to those of males. At Gombe, and probably Mahale and Kibale, a few very peripheral females seem to follow a different strategy and may even maintain relationships with two communities, but these are a small minority. So in contrast to the male-only community hypothesis, most females exhibit spatial allegiance to a male group, and by expanding their territory, males do not immediately "capture" more females.

What then, is the purpose of a male territoriality whereby males repel not only males, but also females of other communities? In agreement with a suggestion by Goodall, Williams and I have found evidence that the males are defending a feeding territory for all the community members. Over the last 20 years the community range of the habituated community at Gombe has fluctuated in area from 8 to 15 square kilometers, apparently because of changes in the strength and territory size of neighboring communities. During times when the range was small, the females had smaller core areas, chimpanzee density was higher, all the chimpanzees were lighter in body weight, and the females had longer interbirth intervals. Therefore, by expanding the community range and thereby excluding female as well as male feeding competitors, the males increased the reproductive rates of the resident females. If the community range remains large, the number of females may eventually increase as dispersing females settle in the area, but females do not seem to be "captured" in the course of the initial expansion as suggested by the male-only community hypothesis.

Cases in which males cooperate to defend a group or territory are rare among mammals in general. More often, cooperation in intergroup conflicts involves groups of females, and sometimes the males associated with them. Male cooperative defense occurs in social carnivores such as lions, cheetahs, and some canids. Among primates, other examples of species in which groups of males cooperate in the defense of females or territory include red colobus, spider monkeys, and some howler monkeys. Mountain gorilla groups usually contain only one male, but where two males are present, both have been observed to cooperate in the violent intergroup interactions characteristic of this species.

Inbreeding Avoidance and Dispersal

During my study of adolescence at Gombe in the early 1970s, I watched the eight-year-old male Goblin become interested in the adult males and follow them. At first he whimpered and became upset if his mother would not stay with him, but gradually he spent more and more time away from her. He followed the alpha male, Figan, who tolerated his presence and supported him in his dominance struggles—first against females, then against males. But when Goblin was fifteen, Goodall observed him turn on his long-term ally and oust him from the alpha position. We now know that, like Goblin, virtually all young male chimpanzees remain in the group in which they are born and eventually join with the other males to defend the community range.

For females it is a different story. At Gombe they remain with their mothers until they are about ten years old, when they start their sexual cycles. Initially they mate with most of the males in the community. However, I found that females with elder brothers or other close male associates ceased to travel with them, and rarely if ever mated with them. Even if the male showed sexual interest in the female, she would scream and avoid him. In addition, females were often fearful of older males.

During my study, one female from the Kasekela and one from the Kahama community reached adolescence. Each, after mating within her natal community, visited the other community while sexually receptive. While both were fearful of and reluctant to mate with some of the males of their own community, they eagerly approached and mated with males from the new community, and these males also appeared to find them more attractive than their own females. While the Kahama female eventually joined the Kasekela community permanently, the Kasekela female returned, pregnant, after six months. Since then, almost all females have made similar visits to other communities during adolescence, and over half of them have left their home community permanently. Over the same period many new females have joined the community and settled there at adolescence; others have visited for various periods, often in the face of considerable aggression from resident females. At Mahale, Taï, and Kibale an even higher proportion of females (about 90 percent or more) have emigrated permanently during adolescence.

The benefit that females gain from avoiding relatives is avoidance of the costs of inbreeding depression, the reduction in the viability and fertility of offspring resulting from mating between relatives.[1] The proximate mechanism driving their behavior appears to be a reduced sexual attraction to individuals with whom they were familiar during immaturity. A similar pattern of reduced sexual attraction to childhood associates was first hypothesized in humans by Westermarck in 1891 and is called the Westermarck effect. It has since been demonstrated in a wide variety of animals.[2]

Patterns of sex-biased dispersal as shown by chimpanzees, in which most or all members of one sex remain in their natal group or area while those of the other sex leave before breeding, are widespread among animals and are viewed by many biologists as an inbreeding avoidance mechanism. In some species, members of both sexes disperse at high rates, and factors in addition to inbreeding avoidance are likely to be involved. In only a few species of mammals such as some whales does neither sex disperse; in these, inbreeding is avoided by mating outside the group. The pattern of male philopatry and female dispersal shown by chimpanzees is, however, unusual among mammals in general and primates in particular. The more common pattern is for most females to remain in their natal area and for most males to disperse (and perhaps in primates, as Strier argues in Chapter 3, for both sexes to disperse). Besides occurring in chimpanzees' close relatives, bonobos, male philopatry occurs in spider monkeys and their close relatives, muriquis, red colobus monkeys, and hamadryas baboons. On the principle of parsimony, therefore, male philopatry has probably evolved four times in primates.

It is likely that chimpanzees show male philopatry because of the unusual advantages that males gain from cooperative territory defense, itself a rare form of behavior among mammals. By staying in the natal group, male chimpanzees will be joining male relatives—fathers, brothers, uncles, and cousins. Cooperation among kin has particular advantages, because even if a cooperating male does not father offspring himself, his assistance will increase the chances that his genes are passed on through the offspring of his relatives, a process known as kin selection. Even in the absence of kin selection, given the nature of competition between groups in which the number of males counts, it would never be possible for a single male to strike off from his natal group and defend a territory on his own. If he had a cohort of males of the same age, he could conceivably do so, as is the case with lions. But chimpanzees reproduce slowly, giving birth to single young rather than litters, and large cohorts of males of the same age never occur. Faced with the presence of their male relatives in the group, females must disperse to other groups to avoid inbreeding.[3]

When we focus on emigration from and immigration into a single group, it is difficult to determine how extensive the movements of female chimpanzees are, and what effect those movements have. Genetic evidence has revealed that gene flow resulting from this behavior is extensive. By collecting shed hair from the night nests of chimpanzees in a variety of sites in East Africa, Philip Morin, David Woodruff, and their associates from the University of California at San Diego have examined patterns in the distribution of mitochondrial DNA. Since mitochondria are usually transmitted only by the female, female residence patterns have a strong effect on the distribution of mitochondrial genes. If females remain close to the place where they are born, then individuals in that area are likely to have similar mitochondrial DNA because they will have inherited it by common descent. If, on the other hand, females migrate and breed in new areas, the mitochondria in a particular area will have a variety of origins and are less likely to be similar. Morin's team found that this was indeed the case. Some individuals at Gombe shared mitochondrial genes with individuals several hundred kilometers away—evidence for extensive successful migration by females in the recent past.

Kinship and Behavior

The consequence of sex-biased dispersal in group-living species is that members of the philopatric sex will be more closely related to one another than the group members of the dispersing sex—provided that mating occurs only within the group, and that members of the dispersing sex do not disperse with relatives. These kinship patterns are widely believed to explain the form of the social bonds within each sex in primates. As W. D. Hamilton pointed out in the 1960s, because individuals can pass on their genes not only by reproducing themselves but also by helping their relatives reproduce, we expect that relatives will often show more helpful or friendly behavior and less harmful behavior to one another than to nonrelatives.

This pattern certainly occurs in many Old World monkeys such as the Gombe baboons; all females stay at home, but all males leave their natal groups singly and join another group before breeding. Females are ranked in strict dominance hierarchies. Close relatives have cohesive, friendly relationships, grooming one another frequently and supporting one another in aggressive interactions against more distant relatives. Nonrelatives sometimes support one another against lower-ranking females, and all the females in the group occasionally support one another against other troops. The unrelated males of the troop also form strict dominance hierarchies, but have tense rather than friendly relationships and do not groom one another.

The rather different kinds of within-sex social relationships found in chimpanzees are congruent both with males being more closely related than females (as we expect from their dispersal patterns) and with their reliance on group territorial defense. Males strive for high rank and form a dominance hierarchy in which there is usually an unambiguous alpha male plus other males ranked as high, middle, or low. However, a male chimpanzee's position in the hierarchy often depends much more on his cooperative alliances with other males than is the case with male baboons. Also, although many social interactions are aggressive and directed at the acquisition or maintenance of dominance rank, male chimpanzees are friendlier to one another than are male baboons. Males groom one another at high rates and spend much time in peaceful association. They have many mechanisms for social appeasement, reassurance, and reconciliation. Frans de Waal has linked the high rates of reconciliation he observed among males partly to the need for unity and cooperation in the face of the external threat of neighboring males. William Wallauer, who has been videotaping the Gombe chimpanzees for the last few years, has some striking film of rivals who never normally groom each other engaged in intense grooming sessions just before they set out to patrol a border area.

Because dominance ranks and dominance behavior are harder to detect among female chimpanzees than among males or among female Old World monkeys, they have received little attention. It came as a surprise, therefore, when Williams, Goodall, and I showed that dominance rank in the Gombe females actually has strong effects on reproductive success. High-ranking females tend to live longer, their infants survive better, they produce surviving infants at higher rates, and their daughters mature as much as four years earlier than those of low-ranking females. Differences in maturation of young suggest that high-ranking families gain better access to food. Possibly high-ranking females are able to occupy core areas containing high-quality food, or they gain preferential access to food when they encounter other females. A more sinister reason for the difference in infant survival is that high-ranking females have been known to kill newborn infants of lower-ranking females.[4]

Taken together, observations of aggression to immigrant females, strong effects of dominance, and infanticide all suggest that competition is intense among females. We are forced to modify the view that females completely avoid competition by feeding alone. Extreme aggression in chimpanzees but not in species such as baboons is clearly congruent with the fact that female chimpanzees are not close relatives. However, female infanticide of close relatives has been observed in nonprimate species such as wild dogs and ground squirrels. It has not been observed in bonobos, a species in which the females are also unrelated. Another reason for aggression at Gombe may be that because of their solitary behavior, female chimpanzees are more vulnerable to attack by coalitions of females than females who are always surrounded by their relatives and allies.

Several studies have started to measure genetic relationships within the community and have raised some questions about the assumption that cooperation among chimpanzees is actually based on kinship. Two analyses have looked at the genetic basis of male alliances within groups, while two others have measured the degree of relatedness of males versus females within communities. Original support for the importance of kinship for male alliances among males of the same community came from observations at Gombe of very close relationships between pairs of maternal brothers. Goodall has described how a male, Faben, initially dominated his younger brother, Figan, and they did not have a very close relationship. After one of his arms was paralyzed by polio, however, Faben lost his rank to Figan in 1966. Faben became a staunch ally of Figan, supporting him in his accession to the alpha position and continuing to side with him until he disappeared in 1974.

I documented a close relationship between another pair of brothers. For several years, adolescent male Sherry followed and groomed his elder brother Jomeo, and Jomeo supported him as he rose in rank against the adult females. Since then, four other close relationships between adolescent males and their elder brothers at Gombe have been observed. Still, once the younger brother reaches full size, the relationship can change. After Sherry defeated Jomeo in a dominance struggle, Goodall found that their relationship waned; and for several years at Gombe, the brothers Freud and Frodo were archrivals for the alpha position.

At Kibale, Harvard anthropologists Goldberg and Wrangham used similarity of mitochondrial DNA to test whether maternal relatives (who would share identical mitochondrial genes by common descent from the mother) formed close relationships. Contrary to expectations, they found that males who had the closest relationships as revealed by grooming and proximity were not maternal relatives, and that individuals sharing the same mitochondrial DNA did not have close relationships. Similar findings were made by the Michigan anthropologist John Mitani and his associates in another community in Kibale. Long-term observation of communities at Gombe, Mahale, and the Arnhem Zoo have also shown that males known not to be brothers can nevertheless form extremely strong alliances. It appears that rather than staying with relatives, adult males are opportunistic in their relationships, making and breaking alliances for individual advantage as the relative power of each male waxes and wanes.

Can we explain the friendlier behavior of males than females on the basis of differences in relatedness? The evidence is mixed. In their 1994 study, Morin and his associates used nuclear DNA extracted from hair to measure the degree of relatedness between males and between females of the Kasekela community at Gombe. They found that males were indeed more closely related to one another than were females within the same community, but the difference was small. A similar study of the levels of relatedness among the Taï chimpanzees by Gagneaux, Boesch, and their associates from the same laboratory found no difference between the relatedness of males and females (see Figure 1.3).

What then should we conclude about the importance of kinship for sex

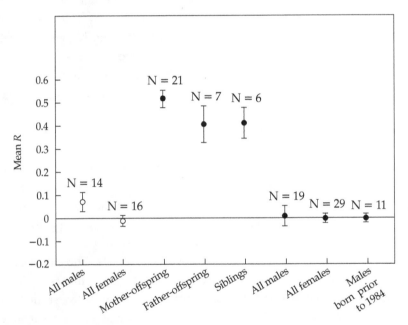

Figure 1.3 Levels of relatedness between different classes of individuals in the communities of Taï (solid circles) and Gombe (open circles), calculated by comparing the genotypes of individuals. The Taï data are based on genotypes at 11 microsatellite loci, the Gombe data at 8 loci. The mean R of zero is calculated by randomly resampling and comparing individuals from the total sample. The background is thus influenced by levels of relatedness among the total group. The mean coefficient of relatedness among mother-offspring pairs is 0.48, which does not differ significantly from the expected value of 0.5. The lower value of 0.36 among maternal sibling pairs probably reflects the inclusion of half-sibling pairs. While males at Gombe are significantly more closely related than females, this is not true at Taï. (Reprinted from Gagneaux et al., 1999, p. 23, © 1999; used with permission of Academic Press.)

differences in cooperation? First, the sample size is very small, and techniques for accurate genotyping are still improving.[5] We need data from a much larger number of communities in order to get a more accurate picture of levels of kinship within communities. Second, it is possible that while patterns of philopatry and dispersal initially evolved because of advantages gained by cooperating, levels of relatedness may be reduced by philopatric kin and strategies of other individuals in the group, such as mate choice by females.

Mating Patterns

The mating patterns of individuals in a group will have strong effects on patterns of relatedness. If a female mates with the same male for successive conceptions, as often happens with humans, her offspring will be full siblings. If females of the same group all mate with the same highest-ranking male, as happens in red howlers and some baboon groups, the young will form paternal sibships. Chimpanzees have complex mating patterns, whose consequences for patterns of relatedness are only just starting to be unraveled.

In theory, in order to conceive a female need mate only once with a fertile male. This is indeed what occurs in some species of mammals like bison. Mating is also infrequent and restricted to fertile periods in monogamous species such as gibbons. But female chimpanzees, like other species living in multimale groups, have adaptations that ensure that they mate hundreds of times before their first conception, and a lower but still significant number of times before the conception of each subsequent infant. Their sexual swellings advertise their readiness to mate. The naked pink skin of the swelling is conspicuous over large distances in the darkness of the forest. Copulations usually take place in full view of the group and last about seven seconds. Copulating females frequently emit a distinctive high-pitched squeal, thus advertising their condition by sound as well as sight (see Figure 1.4). Swollen females mate, on average six to eight times per day, and Goodall observed one who mated fifty times in one day. They are often eager to respond to the mating invitations of all males in the group. Females also mate in the first few months of pregnancy, when they cannot conceive.

Goodall and the Japanese primatologist Akiko Matsumoto-Oda have pointed out several costs of this pattern. When swollen, the female is the center of attention, she is more subject to attack and wounding, and her feeding time is likely to be reduced. There is some physiological cost to producing a large fluid-filled swelling, and the prominent swelling itself is more vulnerable to wounding and infection than areas protected by hair. According to Robert Martin, because mating occurs for several days before and after ovulation, there is also some risk that fertilization will occur between aging (and thus less viable) eggs or sperm. Given these obvious costs, the advantages must be considerable.

While it is generally agreed that the adaptations ensure that females attract and mate with many males, active debate ensues among both primatologists and biologists over how females benefit. Benefits may be material or genetic. In several species, a male modifies his behavior toward a female or her young after mating with her, presumably because this is a reliable indication that he may be the father of her subsequent offspring. For example, each male hedge sparrow in a polyandrous union with a female is more likely to provision the female's offspring and not harm her eggs if he has mated with her. In several species of rodents in which males often kill the young of females, the act of mating inhibits this behavior and sometimes causes the male to show parental behavior toward the young.

Among primates such as Barbary macaques, in which males show helpful behavior toward young, females may mate with many males in order to garner male care. This is probably not a driving force for multiple mating in chimpanzees, because males do not usually participate in much infant care. Females may, however, gain protection from aggression to their infants by mating with all the males.

In the 1970s, following an initial observation by Yukimaru Sugiyama, Sarah Hrdy noted that new males who had just taken over a troop of langur monkeys and evicted the previous male, often killed the young infants present in the group. The effect was to speed up the female's return to sexual receptivity so that the male was able to mate with the female and father an infant more quickly. This behavior has since been observed in a number of other primates, as well as in other mammals such as lions, and in rodents, and Hrdy's adaptive explanation is generally supported. She noted that female langurs showed situational "estrus behavior" in which they mated with new males even when pregnant, and that females might mate with many males to confuse paternity, thus forestalling infanticidal or other harmful behavior of the male toward the female.

Infanticide by male chimpanzees has been observed at Gombe, Mahale,

A.

Figure 1.4 A. An adult male chimpanzee in a typical sexual solicitation pose, inviting a female. B. A young adult female with genital swelling. C. A female screams at the climax of copulation. (Photos by Frans de Waal.)

B.

C.

and in Budongo, Uganda. All cases of infanticide at Gombe have involved males killing the infants of females from other communities. However, the males were not subsequently seen mating with the females, and this behavior is perhaps best interpreted as a manifestation of general hostility to neighbors who are food competitors. At Mahale, however, males have killed several infants of females within their community. All victims were the first- or second-born infants of newly immigrated females, and all were male. In most of these cases the infanticidal male had mated with the female, but these females were part of an unusual influx of females to the M-group from the K-group after all but one of the adult males had disappeared. The females continued to spend much time at the edge of or outside the range of the new community, away from the males. Reviewing these cases, Mariko Hiraiwa-Hasegawa and Toshikazu Hasegawa suggest that a female's ranging patterns and the amount of time spent with males can be used as a cue to determine her infant's paternity; by their criteria, these infants were of dubious paternity. After the infanticides the females gradually spent more time with the males, and their subsequent infants were not killed.

It is likely, then, that female chimpanzees protect themselves from male infanticide by attempting to associate and mate frequently with males. This pattern may explain the tendency documented by reproductive biologist Janette Wallis of younger females to cycle longer into pregnancy, and to resume cycling more quickly after the birth of their babies. Such females are likely to be less familiar to the males and may benefit from frequent interaction with them. Other material benefits that females may receive from males in exchange for mating include general support and protection from other chimpanzees, and increased access to food under the control of the male.

The idea that females can gain significant genetic benefits from mating with multiple males has been gaining acceptance. If males vary in the quality of their genes, for example in terms of their effects on survival or resistance to disease, females will profit by mating with the highest-quality male to gain these genes for their offspring. By attracting many males to them, they may more easily be able to select the best male, or just passively incite competition among males so that the most competitive male mates with them. Mate selection and competition can continue in important ways inside the female after mating. As biologists Jeanne Zeh and David Zeh note in a fascinating review, sperm do not face an easy ride inside animals, including humans, subsequent to deposit in the vagina. "Sperm are perceived as antigens and must run the gauntlet of a female reproductive tract populated by large numbers of anti-sperm leucocytes and antibodies. Of the 40–1800 million sperm deposited, for example, in the human vagina only about 300 reach the site of fertilization." If sperm of more than one male are present at the same time, there will be competition to reach the ovum. Zeh and Zeh also suggest various reasons why there may be different levels of genetic compatibility between the genomes of different mates, so that male quality or sperm quality is a relative characteristic. Thus females may benefit from sampling many males, and different females will not necessarily benefit from mating with the same "high-quality" male.

Given the tendency of females to seek and mate with multiple males, what do males do? Goodall and her associates Patrick McGinnis and Caroline Tutin have described three mating patterns at Gombe, which are also observed at the other sites. In opportunistic mating, all males in the group may mate with the female and there is little overt competition. Often, however, and particularly toward the end of the swelling period when ovulation is more likely, one male, usually the most high-ranking male in the party, shows possessive behavior. He remains close to the female, and threatens or attacks other males and often the female if they attempt to mate. The Yale anthropologist David Watts found that in an exceptionally large community of chimpanzees at Ngogo in Kibale, the success of single males in guarding the female decreased as the number of males in the party increased. When male party size exceeded about 14, the top-ranking males started to cooperate to guard the female, sharing both copulations with her and the costs of keeping other males away. A third pattern is when a male takes the female on a consortship and sequesters her on the periphery of the community range for a period of days, and occasionally weeks or even months. On consortships the female benefits from less feeding competition and less stress, but the pair is at increased risk of attack from neighbors. In some consortships females follow males willingly, but in others the male aggressively coerces the female to accompany him. Attacks of cycling females by males are quite frequent, and Goodall suggests that in general these force the female to be compliant in sexual situations.

In addition to these behavioral traits of aggressive competition, mate guarding, and sexual coercion, competition between males over females has selected for several physical traits in males. As in other species in which there is competition among many males for access to receptive females, male chimpanzees are larger than females. Charles Darwin was the first to point out that sexual dimorphism in size and weaponry is likely to result from selection on males to compete in aggressive competition over females. Selection has also acted on genital morphology. Because a female often mates with many males, her reproductive tract is apt to house sperm from more than one male. Therefore, the male who is able to deposit the most sperm is the most likely to gain a successful fertilization. Sperm number depends on size of testes, and male chimpanzees have enormous testes compared to those of other ape species.[6] In many species in which there is a potential for sperm competition, other adaptations have evolved to prevent successful insemination by other males. For instance, the ejaculate of male chimpanzees and males of species such as baboons forms a glutinous plug in the female's vagina that impedes subsequent insemination by other males.

Who Are the Fathers?

With these conflicting strategies of females and males for achieving mating, which males actually father the young? The answer may be determined in two ways: by examining behavioral data on mating during periods in which conception occurred, and by genetic studies. Behavioral studies have reached different conclusions about the success of dominant males in monopolizing females. In the Mahale Mountains, consortships are uncommon; most conceptions are thought to occur in the group-mating situation; and alpha males are often possessive of females, suggesting that male rank is likely to influence male reproductive success.

At Gombe, consortships are quite common. Estimates of the proportion of conceptions that have taken place on consortships vary from 50 percent in a 1981 study by Tutin and McGinnis, to 25 percent in later studies by Goodall and Wallis. Several different males were involved in consortships that were thought to lead to conception, and although some males were high ranking, only a small proportion ever reached alpha status. Goodall estimated that up to 15 percent of conceptions took place during visits of females to other communities, but the majority appeared to occur during group-mating situations within the community. While alpha males sometimes attempted to monopolize females, fertile females often mated with several males, even during the days at the end of the swelling when the female was most likely to ovulate. Goodall concluded that there was little relation between male rank and reproductive success, and that a variety of males achieved paternity.

Genetic studies of the Gombe chimpanzees are still in progress. In 1994 Morin and his colleagues genotyped all the individuals in Gombe's Kasekela community from their hair. However, they were able to assign paternity in only two cases, because many of the potential fathers of the current crop of infants had died before their hair was collected. Currently, Julie Constable is genotyping additional infants from hair and feces. The data so far suggest that while about 20 percent of conceptions occurred during consortships with mid- or low-ranking males, the majority occurred during group mating within the group. Most males who achieved fertilization were either alpha male at the time or became alpha male at some point in their lives. Their success occurred despite the fact that they did not prevent all other males from mating with the female during the cycle. Three females in the sample had more than one infant, and in all cases the fathers were different males.

In the Taï Forest, Gagneaux and his colleagues genotyped fourteen infants, their mothers, and potential fathers. They were able to assign paternity of seven infants to six different males. While the fathers of four were alpha males at some point in their lives, only two were alpha male during the period of conception. The most striking finding was that seven of the infants may not have been fathered by any of the community males. Instead, the researchers suggest, the mother must have left the community and mated with a male in another community. However, techniques for obtaining accurate genotypes are still improving, and the Gagneaux group is currently working to confirm these results. At Bossou, where the study community was very small and only included one adult male, Sugiyama and his associates found evidence that one infant may have been conceived outside the community.

Preliminary as they are, these data suggest that the relationship between male rank and reproductive success is strong but not absolute. Females are quite successful in mating with a variety of males and possibly even risk leaving the safety of the community to do so. The result of these mating patterns is that males within chimpanzee communities are not as closely related as they could be. Females mate with so many males that brothers who are full siblings probably occur only rarely. Because of low birth rates, turnover of alpha males, and possible mating outside the group, cohorts of paternal siblings resulting from females mating with the same alpha male (as in species such as the Amboseli baboons) may also be rare.

Human and Chimpanzee Social Structures

Some strong similarities exist between human and chimpanzee social structure. Social groups in human hunter-gatherer societies often consist of about 150 members, not much larger than the largest chimpanzee communities. Human societies also resemble chimpanzees in patterns of residence. According to Lars Rodseth and his associates, human societies show a variety of residence patterns but "a substantial majority of human societies are characterized by female-biased dispersal and male philopatry." This paradigm is particularly evident in agricultural societies, but even among hunter-gatherer societies almost two-thirds show male philopatry, while less than a fifth show female philopatry.

The intergroup hostilities of chimpanzees also resemble those of some human societies. The noisy posturing at each other of groups of similar size resembles the ritualized "bluffing warfare" observed in some traditional societies. Manson and Wrangham have drawn attention particularly to the similarity of the patrols of chimpanzees that end in the tracking down and attacking of lone individuals, to the lethal raiding observed in many human societies. However, the behavior of chimpanzees falls short of more organized human warfare involving pitched battles between groups of opponents that result in many casualties.

Given the rarity of male philopatry and male cooperation in intergroup hostility among animals, it is striking that these patterns occur in our closest relatives, chimpanzees, and to some extent in bonobos and gorillas. It is likely, therefore, that our common ancestor also shared this pattern. The importance of kinship in human social relationships has always been evident to anthropologists. Cooperative male sibships are prominent in many traditional societies. But as Rodseth and his associates point out, a significant way in which human societies differ from those of chimpanzees is that kin bonds have become elaborated and maintained even beyond the boundaries of the group, in part by exchange of females in marriage.

The most striking difference between chimpanzee and human social structure lies in mating patterns and sexual behavior. In humans, the broad pattern is for conjugal units to exist within the larger social group. Long-term pair bonds between a male and female are universal, although their form, strength, and duration vary between societies, and in many societies some males may be bonded simultaneously to more than one female. Unlike those of chimpanzees and other group-living primates, human copulations are private. Consequently it is difficult to get accurate estimates of either mating frequency or pair fidelity. It is evident that humans mate at lower frequencies than receptive female chimpanzees, with the majority mating only once every few days (according to most surveys). On the other hand, humans mate over more of their total life span than chimpanzees. Mating is more prolonged and diverse in form and may occur throughout the menstrual cycle, pregnancy, lactation, and after reproductive cessation in females. Although most societies carry severe sanctions against adultery, extra-pair conceptions do occur in both traditional and modern societies (for example, from the records of one maternity hospital in North America, 10 percent of conceptions were extramarital), but levels of promiscuity are clearly lower than in chimpanzees.

How do human secondary sexual characteristics relate to these patterns? Human males are about 15 percent larger and more muscular than females, a degree of sexual dimorphism that is somewhat lower than the 25 percent in chimpanzees but considerably more than that in such monogamous species as gibbons, in which sexes are the same size. In contrast to these fairly comparable levels of size dimorphism, humans have very different patterns of genital dimorphism than chimpanzees (Figure 1.5). Relative size of testes is much smaller and is comparable to that of monogamous species or those living in one-male groups.

Unlike chimpanzees, human females do not show the cyclic changes in the genital area that advertise ovulation, and they are the only apes—indeed, the only primates—to show a conspicuous swelling of the breasts coincident with the onset of sexual maturation rather than lactation. The breasts are interpreted by many to be a continuous sexual signal. "Continuous" sexual advertisement and sexual receptivity of females are seen by some as adaptations to retain the interest and loyalty of the mate. They do so either by keeping the male sexually satisfied or, as Katherine Noonan and Richard Alexander suggest, by threatening the male with infidelity if he does not return to the female frequently. Hrdy and others see these adaptations as a way for females who are generally constrained by mate guarding to engage in sexual behavior with other males when a situation presents itself and it is advantageous to the female to participate.

The critical question for understanding the evolution of human social structure is, How can we explain the evolution of the human pair bond in which both females and males have forgone high rates of promiscuity? Two major hypotheses exist, and they are also taken up in other chapters of this book. Until rather recently, most people subscribed to the view that pair bonds evolved to facilitate the exchange of resources between the sexes. In particular, this theory has been tied to an increased reliance on meat, which is procured by males. As division of labor evolved with females gathering vegetable food while males hunted, both sexes benefited by sharing. Some scenarios also point to the increasingly long period of helplessness and dependency of human infants as brain size increased,

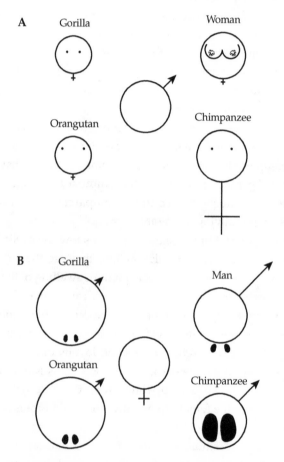

Figure 1.5 A. The male's view of the female in different great-ape species, indicating the degree of sexual dimorphism in body size (relative size of the circles), and the relative development of the mammary glands and swelling of the perineum (cross) prior to first pregnancy. B. The female's view of the male, indicating the degree of sexual dimorphism in body size (circle size), the size and position of the testes (black), and the relative size of the erect penis (arrow). (Modified from Short, 1979.)

and suggest that males began to help females with infant care and protection, as is observed in some monogamous primates.

Barbara Smuts and others have proposed an alternative hypothesis that emphasizes the advantages that females receive from their mate in protection against sexual coercion by other males. She points to the existence of long-term special relationships between male and female savanna baboons in which the female apparently benefits from male protection and investment in her infant, while the male may benefit from increased mating opportunities with the female. She suggests that males in an initially chimpanzee-like society might have evolved to respect these bonds between even low-ranking males and their mates, if male cooperation became so important that it was worthwhile to tolerate such bonds in return for the male's cooperation. If so, we need to explain the factors that selected for such an increase in male cooperation during human evolution. In a variant of this "bodyguard" idea, Wrangham (see Chapter 5) argues that after the invention of cooking, females needed males to guard their food.

Although both ideas are based on changes in diet and way of life that occurred after our ancestors left the trees and inhabited the forest edge and savanna, further study of chimpanzees may yield insights into factors facilitating pair bonding. Examples include the occasional occurrence of more prolonged relationships between pairs of males and females, as have recently been reported in Taï, the contexts of food-sharing and possible barter for sexual and other favors (see Chapter 4), and the contexts of male cooperation and toleration of the mating of subordinates. Unfortunately, opportunities to study wild chimpanzees are dwindling rapidly with increased pressures from habitat destruction and decimation of chimpanzee populations because of the bushmeat trade. Already in many places, habitat fragmentation and isolation of chimpanzee communities are curtailing the opportunities for females to mate widely, which will have consequences for genetic diversity. We have both a moral duty to our closest relatives and an urgent need as scientists to aid in their conservation.

2

Apes from Venus: Bonobos and Human Social Evolution

Frans B. M. de Waal

Even though bonobos are as close to us as chimpanzees, the species is less well known both to science and to the general public. These "make love, not war" primates have evolved peaceful societies with female bonding and female dominance. As such, bonobos challenge traditional assumptions about human social evolution.

The bonobo is an extraordinarily sensitive, gentle creature,
far removed from the demoniacal Urkraft [primitive force]
of the adult chimpanzee.
—*Tratz and Heck, 1954*

HAD THE BONOBO been known to science first and the chimpanzee second, we might today have different ideas about the inevitability of violence in human society, about male dominance and male bonding in hunting and warfare, about the role of technology, and about the social significance of sex. Inasmuch as the bonobo fails to support male-biased evolutionary scenarios (summarized in dramatic titles such as *Man the Hunter, Man the Tool-Maker, Men in Groups, Demonic Males, The Imperial Animal,* and so on), these scenarios might not have become nearly as popular as they are. Not that the bonobo would have inspired more accurate reconstructions of the human past than the chimpanzee—at this point we are unable to tell—but in the bonobo we do have a female-centered and relatively peaceful close relative.

In the late 1990s many strands of knowledge came together concerning this relatively unknown anthropoid ape. The findings deserve serious attention. I review here what we know and do not know about the bonobo, then speculate about how its behavior enriches models of human social evolution.

Discovery of the Bonobo

The historic discovery of the species took place in a colonial Belgian museum following the inspection of a skull by a German anatomist, Ernst Schwarz, who in 1929 claimed to have found a new subspecies of chimpanzee. Soon thereafter, the morphological differences were considered important enough to separate the bonobo as a full-fledged species. It was placed within the same genus, *Pan*, as the chimpanzee. Despite the fact that the bonobo, *Pan paniscus*, came to be known as a dwarf or pygmy chimpanzee, those names have now fallen out of favor. One reason is that with an average weight in captivity of 43 kilograms for adult males and 37 kilograms for adult females, bonobos reach about the same size as the smallest subspecies of chimpanzee.

The bonobo has reddish lips in a black face, small ears, and wide nostrils. The species also has a flatter, more open face with a higher forehead than the chimpanzee, and an attractive coiffure with long, fine, black hair neatly parted in the middle (Figure 2.1). The overlap in weight and size notwithstanding, the bonobo has a more elegant build than the chimpanzee, with strikingly elongated legs. The result is that, when knuckle-walking on all four limbs, the bonobo's back remains relatively horizontal because of the elevated hips, whereas the chimpanzee's back slopes down from the powerful shoulders. When standing or walking upright—which the bonobo often does, especially when carrying food—the bonobo's back seems to straighten better than that of the other apes, giving it a strikingly humanlike posture (Figure 2.2).

Harold Coolidge, the anatomist responsible for the bonobo's species status, considered this ape anatomically more generalized than the chimpanzee, noting in 1933 that the bonobo "may approach more closely to the common ancestor of chimpanzees and man than does any living chimpanzee." Following up on this suggestion, Adrienne Zihlman measured the body-weight distribution in all members of the hominoid family, especially the relative weights of arms and legs. She found that in this regard the long-legged bonobo most resembled the australopithecines.

Her findings do not mean that the bonobo is our nearest relative or that *Homo sapiens* descended from the bonobo. That would be impossible, in that the split between the two *Pan* species occurred millions of years after we separated from them. But perhaps the bonobo's body plan has undergone relatively few modifications since the mysterious common ancestor walked the earth—an estimated 6 million years ago. Our own lineage evolved dramatic new characteristics: bipedal locomotion, a large brain, and bare skin. In the chimpanzee, evolutionary changes may have been prompted by a need to adapt to half-open habitats such as woodlands. The bonobo, on the other hand, may never have left the protection of the rain forest: at present, the species is almost entirely restricted to wet equatorial regions. As a result, it may retain the largest number of ancestral traits.[1]

Current debates are therefore not about which species is closer to us, the bonobo or chimpanzee, but rather which one more resembles the last common ancestor. Despite passionate claims both ways, the matter is far from decided. The two outliers to look at when comparing the chimpanzee and the bonobo are the gorilla and our own species. With the gorilla, the bonobo shares thick-walled nostrils and dark-colored infants (as opposed to the light-faced chimpanzee infants). With us, the bonobo shares, apart from limb proportions, a longer gestation period and a more humanlike distribution of special neurons in the frontal cortex. This last finding illustrates the sort of data that one day may settle the issue.[2] Because a similar list of characteristics shared with outlier species can be produced for the chimpanzee (see Chapter 5), it is too early to choose between the two. Preferences risk being based either on wishful thinking (some people regard bonobo characteristics as desirable) or on commitment to traditional scenarios of human evolution from before we knew much about the bonobo. Inasmuch as existing scenarios have been disproportionately affected by

Figure 2.1 Portraits of a bonobo (left) and a chimpanzee (right). Both are fully grown males. Note the bonobo's gorillalike thick-walled nostrils, his longer, finer hair, and his light-colored lips compared to the chimpanzee. (Photos by Frans de Waal.)

Figure 2.2 Bonobos standing upright, like this female (left) and male at the San Diego Zoo, resemble an artist's impression of *Australopithecus*. In terms of leg length relative to the rest of the body, bonobos are more humanlike than any other great ape. They are often observed to walk bipedally, especially when carrying food. (Photo by Frans de Waal.)

knowledge of only one close relative—the chimpanzee—we should not be surprised if the behavior of this relative best fits them.

Personally, I prefer the default position—which is that both *Pan* species tell us something about the common ancestor, and that we therefore should consider them equally in the challenging task of reconstructing the past.

The Killer Ape Myth

In the 1960s and 1970s a much more distant relative, the savanna baboon, was widely regarded as the best model of the human ancestor (see Chapter 3). This terrestrial primate is adapted to the sort of ecology that proto-hominids were thought to have faced after descending from the trees. The baboon model was abandoned, however, when it became clear that fundamental human characteristics are absent or only minimally developed in this monkey, yet present in the chimpanzee. Cooperative hunting, food sharing, tool use, power politics, and primitive warfare were observed in chimpanzees. In the laboratory these apes learned symbolic communication, such as sign language, and recognized themselves in a mirror—achievements monkeys can not attain. Chimpanzees are, of course, also much closer to us genetically than are baboons.

The chimpanzee as the touchstone of hominid evolution played a critical role in settling the debate about the aggressive nature of our species, which started well before we knew much about the chimpanzee in the wild. In 1925 Raymond Dart announced the discovery of *Australopithecus africanus*, a missing link in the human fossil record. This bipedal hominid with ape-like features placed the human lineage very close indeed to that of the apes. On the basis of rather flimsy evidence, Dart speculated that *Australopithecus* had been a bloodthirsty carnivore, an idea dramatized in *African Genesis* by science writer Robert Ardrey. The killer ape myth was born, including a connection between warfare and hunting and the concept that aggressiveness drives cultural progress. In 1963 Konrad Lorenz added, in his influential *On Aggression,* that our species unfortunately has not had time to evolve the same inhibitions that we see in "professional" carnivores, such as lions and wolves. As a result, we are more dangerous to our own kind.

This pessimistic view soon met with resistance. Social psychologists and anthropologists demonstrated that aggressive behavior can be learned, and they questioned the universality of human violence by pointing at cultures believed to be completely peaceful. Yet the premises of this debate were fundamentally flawed. It was tacitly assumed that demonstrating either a genetic or a learning component would settle the issue, whereas we now assume involvement of *both* influences in almost anything humans do. In addition, the critics pointed at the great apes as proof that aggressive instincts cannot be highly developed in our species: in those days, apes were seen as peaceable vegetarians.[3]

Here they made two fundamental errors, which backfired a decade later. First, they tried to have it both ways: denying the influence of biology while at the same time appealing to what was known about apes to make a point about human evolution. Second, the image of apes was to change dramatically in the late 1970s with observations of predation on monkeys, lethal intercommunity aggression, infanticide, and even occasional cannibalism. Because the critics had relied on ape peacefulness as an argument, they could not suddenly deny the relevance of ape aggressiveness. The data on chimpanzees strongly endorsed the earlier scenarios, which as a result took on the aura of absolute truth, most recently restated by Richard Wrangham and Dale Peterson in *Demonic Males:*

> That chimpanzees and humans kill members of neighboring groups of their own species is . . . a startling exception to the normal rule for animals. Add our close genetic relationship to these apes and we face the possibility that intergroup aggression in our two species has a common origin. This idea of a common origin is made more haunting by clues that suggest modern chimpanzees are not merely fellow time-travelers and evolutionary relatives, but surprisingly excellent models of our direct ancestors. It suggests that chimpanzee-like violence preceded and paved the way for human war, making modern humans the dazed survivors of a continuous, 5-million-year habit of lethal aggression.

It is at this point that the bonobo, had it been known before the chimpanzee, would have made a difference. The rate of violence is considerably lower in the bonobo, and the species seems much less keen on animal protein. Thus far, we have no evidence for intercommunity raiding, infanticide, cooperative hunting, or any of the other lethal activities stressed as the hallmark of our species. Whereas the chimpanzee as ancestral model fits the killer ape myth, the bonobo would have played into the hands of the opposition.

It is true that the current way of thinking about human and ape aggression assumes greater flexibility; that is, it has moved away from the Lorenzian drive concept (which made aggression inevitable) and searches instead for environmental determinants. In this view violence is an option, expressed only under special ecological conditions. Nevertheless, since what we know about the bonobo casts doubt on the assumption of pervasive violent tendencies in our evolutionary history, attempts have been made to downplay the importance of this close relative as either an intriguing yet specialized anomaly that can be safely ignored, or as not quite as distinct from the chimpanzee as claimed.

Make Love, Not War

The reason to speak of the killer ape *myth*, instead of theory or concept, is not because humans are angels of peace—they are not—but because it is a narrow scenario that became tremendously popular after the horrors of World War II. Confidence in human nature was at a low, and the view that the human being is a lustful killer—or, as Matt Cartmill summarized the central thesis, "a mentally unbalanced predator, threatening an otherwise harmonious natural realm"—went down remarkably easily with scientists and general public alike.

The myth failed to distinguish between predation and aggression, and did little justice to the complexity of our species and its many distinct features *other* than aggressiveness. It looked at tools as having their origin in weapon use (a position unsupported by the ape evidence), ignored the role of culture and language, ascribed to hunting a far greater contribution to subsistence than it probably ever made, and granted women no role other than as the bearers of children and the passive objects of male competition. In place of this myth we need a scenario that acknowledges and explains the virtual absence of organized warfare among today's human foragers, their egalitarian tendencies, and generosity with information and resources across groups. If our ancestors lived for perhaps millions of years in these kinds of societies, the hierarchical and territorial behavior said to typify our species may well be a relatively recent phenomenon instead of a "5-million-year habit." Similarly, it has been pointed out that today's hunter-gatherers depend more on the gathering activity of women than the hunting activity of men. In other words, the importance of hunting in our heritage may well have been overrated.

Instead of further exploring these criticisms, I will focus on the issue of aggression and its counterpart, the resolution of intragroup conflict. Humans have a truly remarkable ability for peaceful coexistence, amply demonstrated in an ever more crowded and urbanized world. This is part of our primate heritage. We know now that the connection between crowding and aggression, initially demonstrated in rodents, fails to hold in monkeys and apes. Under crowded conditions these animals increase appeasement signals and grooming contact, thus managing social tensions so effectively that in some species aggression levels are virtually independent of population density. These findings counter the message of Ardrey and Lorenz, and hint at the significance in primate evolution of mechanisms that check the destructive potential of competition.

One mechanism that has received much attention is *reconciliation*. First reported in 1979 for the large chimpanzee colony of the Arnhem Zoo in the Netherlands, it is now known in many different species, including even some nonprimates. The principle is simple: shortly after a fight between two individuals, these same individuals come together to engage in positive contact, such as kissing and embracing (in chimpanzees), or handholding, mounting, grooming, or clasping in other species. Instead of dispersal—traditionally considered the chief function of aggressive behavior—we actually see an intensification of friendly behavior subsequent to fights. As reflected in the "reconciliation" label, the hypothetical function of these friendly exchanges is the restoration of long-term relationships. Over the last twenty years or so, evidence for this function based on both observational and experimental research has been mounting.

Reconciliation is more common among bonobos than chimpanzees. Paradoxically, the presence of peacemaking implies aggressiveness: in the absence of conflict there is no need for reconciliation. The mere fact that bonobos show this kind of behavior implies that they, like chimpanzees, use dominance and aggression to settle conflict. The level of violence may be lower in bonobos, and the degree of social cohesiveness greater. It would be grossly inaccurate to depict the species as entirely peaceful: the difference from chimpanzees is slight.

Setting out to study the aftermath of aggression in the bonobos of the San Diego Zoo in California (at the time, the world's largest captive colony), I spent my time in front of the ape enclosure with a video camera. The camera was switched on at feeding time, which turned out to be the peak of sexual activity. As soon as an animal caretaker approached the enclosure with food, the males would develop erections. Even before the food was thrown into the enclosure, the bonobos would be inviting each other for sex: males would invite females, and females would invite males as well as one another.

In almost all animals, the introduction or discovery of attractive food induces competition. There are two reasons to believe that sexual activity is the bonobo's answer to this condition. First, any object, not just food, that arouses the interest of more than one bonobo at a time tends to lead to sexual contact. What such situations share with feeding time is not a high level of excitement, but the possibility of conflict over possession. Whereas the result in most other species is squabbles, bonobos are remarkably tolerant, perhaps because they use sex to divert attention and change the tone of the encounter (Figure 2.3).

Second, bonobo sex also often occurs in aggressive contexts unrelated to food. For example, one male chases another away from a female, after which the two males reunite and engage in scrotal rubbing. Or one female hits a juvenile and the juvenile's mother responds by lunging at the aggressor, immediately followed by sex between the females. Chimpanzees kiss and embrace during reconciliations yet rarely engage in sex, whereas bonobos use the same sexual repertoire then that they do at feeding time.

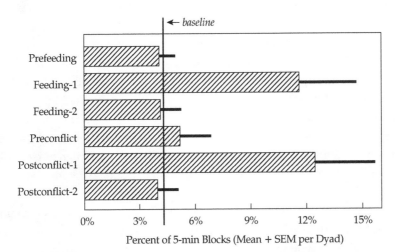

Percent of 5-min Blocks (Mean + SEM per Dyad)

Figure 2.3 Mean (+ SEM) percentage of five-minute blocks of sociosexual behavior among the bonobos of the San Diego Zoo, under six conditions related to food availability (immediately before feeding time, immediately afterward, and 15 minutes after feeding) and spontaneous aggression (immediately before a conflict, immediately afterward, and after a delay of 15 minutes). The graph shows sharp increases in sociosexual activity around feeding time and after aggressive incidents.

Based on an analysis of many such incidents, my study yielded the first solid evidence for sexual behavior as a mechanism to overcome aggression. The art of sexual reconciliation may well have reached an evolutionary peak in the bonobo.

Occasionally the role of sex in relation to food is taken one step further, a step that brings bonobos very close indeed to humans. It has been speculated by anthropologists such as Owen Lovejoy and Helen Fisher that the reason for the partial separation between sex and reproduction in our species is that sex serves to cement mutually profitable relationships between men and women. The human female's capacity to mate throughout her cycle, and her strong sex drive, make it possible to exchange sex for male commitment and paternal care, thus tying men and women together in the nuclear family. Instead of a conscious strategy, this arrangement is thought of as favored by natural selection for the simple reason that it allows women to raise more offspring than they could on their own. Although bonobos clearly do not establish the exclusive heterosexual bonds characteristic of our species, their behavior does fit important elements of this model: female bonobos show extended receptivity (voluntarily engage in sex during much of their menstrual cycle) and use sex to obtain male favors. Thus, a female who does not dominate a particular male still can rely on sex as a weapon.

At the San Diego Zoo I observed that if Loretta was in a sexually attractive state (which she was during a large portion of her cycle) she would not hesitate to approach the adult male Vernon if he possessed food. She would present herself to Vernon, mate with him, and make high-pitched food calls while taking over his entire bundle of branches and leaves. She would hardly give Vernon a chance to pull out a branch for himself, sometimes grabbing the food out of his hands in the midst of intercourse. This was quite a contrast with periods in which Loretta had no genital swelling; then she would wait until Vernon was ready to share.

Suehisa Kuroda reports similar exchanges at the provisioning site maintained by Japanese fieldworkers in the Democratic Republic of Congo (formerly Zaire): "A young female approached a male, who was eating sugar cane. They copulated in short order, whereupon she took one of the two canes held by him and left." In another case, "a young female persistently presented to a male possessor, who ignored her at first, but then copulated with her and shared his sugar cane."

Despite the occasional quid pro quo between the sexes, bonobos do not form humanlike nuclear families with permanent bonds between a male and one or several females; the burden of raising bonobo offspring rests entirely with the female. In fact, nuclear families are probably incompatible with the widespread use of sex by bonobos.

The same within-group use of sexual behavior to smooth tensions seems to extend to relations between groups. The pattern is a distinct contrast to that of chimpanzees, whose males are known to patrol the borders of their territory, occasionally invading a neighbor's territory and causing lethal raids between communities. In bonobos, not a single report of this level of intercommunity violence exists. Peaceable mingling of communities seems to be the rule, including mutual sex and grooming. Friendly overtures are made mostly by females; male-male relations between groups are relatively tense. Perhaps owing to the virtual absence of intergroup violence, males have less need to join forces and maintain close ties. All-male associations are unusual. Most typical is the mixed party, in which adult males, adult females, and youngsters travel together.

Variable Sexuality

The very first suggestions that the sexual behavior of bonobos might differ from that of chimpanzees came from observations at European zoos. Wrapping their shocking findings in Latin, two German scientists, Eduard Tratz and Heinz Heck, reported that chimpanzees mate *more canum* (like dogs) and bonobos *more hominum* (like people). In those days, the 1950s, anthropologists viewed face-to-face copulation as a human cultural innovation, ignoring the fact that our species is anatomically adapted for such intercourse.

These early studies did not reach the international scientific community partly as a result of their publication in the German language, but also because of a tendency to dismiss unusual behavior in zoo animals as an artifact of captivity. Could it be that bonobos act so "grotesquely" because they are, unbearably bored, or because they are under human influence? We know now that except under extreme conditions the effects of captivity on the behavioral repertoire of a species are much less dramatic than formerly assumed. Whatever the conditions under which *other* primates are kept, they never act like bonobos. In other words, it must be the species rather than the environment that produces the bonobo's characteristic behavior. It was only when fieldwork took off that the behavior-as-artifact explanation could be put to rest. Research in the natural habitat has validated rather than contradicted the early observations of Tratz and Heck.

At the provisioned site of Wamba, sexual activity in all partner combinations is seen in relation to food or after social tensions. Similarly, at the unprovisioned site of Lomako Forest sexual activity increases during the sharing of meat or when a group of bonobos enters a fruiting tree. These studies also suggest extended female sexual receptivity. The female's tumescent phase, which results in a conspicuous pink swelling that signals willingness to mate, covers a greater part of the menstrual cycle in bonobos than in chimpanzees, and bonobo females also mate when they are not in the swelling phase. Instead of a few days of her cycle, therefore, the female bonobo is almost continuously sexually active. As Alison and Noel Badrian put it, "Pygmy chimpanzees copulate even when the potential for reproduction is low."

It is impossible, therefore, to understand the social life of the bonobo without attention to its sex life. This proviso has turned out to be a bit of a handicap for the distribution of knowledge. Film crews have traveled all the way to Africa to film bonobos, only to turn off their cameras in embarrassment during sexual scenes; general-interest magazines have strenuously looked for euphemisms (such as "very affectionate") to describe bonobo relations; and some scientists have selectively ignored sex with same-sex partners, discussing only heterosexual sex when comparing bonobos and chimpanzees. Of course, once the news about bonobos did hit the media, the disadvantage of their eroticism quickly turned into an attention grabber. Initially, however, this was not the case, and the species remained obscure far longer than necessary.

Whereas in most other species sexual behavior is a fairly distinct category, in the bonobo it has become part and parcel of social relationships—and not just between males and females (see Figure 2.4). Bonobos engage in sex in virtually every partner combination: male-male, male-female, female-female, male-juvenile, female-juvenile, and so on. They become aroused remarkably easily and express their arousal in a variety of mounting positions and genital contacts. Sexual behavior is flexible: bonobos use every imaginable position and variation. Whereas chimpanzees almost never adopt face-to-face positions, bonobos do so one out of three times during copulations in the wild, and even more often in captivity. Furthermore, the frontal orientation of the vulva and clitoris suggest that the female genitalia are adapted for this position.

Perhaps the bonobo's most characteristic and most common sexual pattern is the so-called genito-genital rubbing, or GG-rubbing, between adult

Figure 2.4 Genito-genital rubbing of adult female bonobos. The participants often bare their teeth and squeal in apparent pleasure. This pattern, the most common in the bonobo's rich sexual repertoire, is politically the most significant. Even when unrelated, bonobo females establish close bonds, which translate into alliances that help them gain powerful positions in society. (Photo by Frans de Waal.)

females (Figure 2.4). One female clings with arms and legs beneath another (almost the way an infant clings to its mother) while the other female, standing on both hands and feet, lifts her off the ground. The two females then rub their genital swellings together laterally with the same rhythm as that of male thrusting during copulation. Given the rather prominent clitoris of bonobo females, there can be little doubt that their grins and squeals during GG-rubbing reflect orgasmlike experiences. Absent in the chimpanzee, this behavior has been observed in every bonobo group, captive or wild, containing more than one female.

Male bonobos, too, may engage in pseudocopulation, but generally perform a brief scrotal rub instead: the two males stand back to back, then one male rubs his scrotum against the behind of the other. Males also perform so-called penis-fencing: in this rare behavior, thus far observed only in the field, two males hang face to face from a branch while rubbing their erect penises together as if crossing swords. The sheer variety of sexual and erotic contacts in bonobos is impressive, especially if we include sporadic oral sex, massage of another individual's genitals, and tongue-kissing.

Could the difference in sexual activity between chimpanzee and bonobo be a by-product of differences in ecology, social grouping, and foraging requirements? The botanist who sees different plants grow at different speeds in different environments has an easy solution to this sort of problem: he moves seedlings of both plants to a greenhouse to compare their growth under identical conditions. Similarly, the question about differences in sexual behavior between the two *Pan* species is best answered by studies under controlled conditions in which both species have the same amount of space, the same kind of food, and the same number of part-

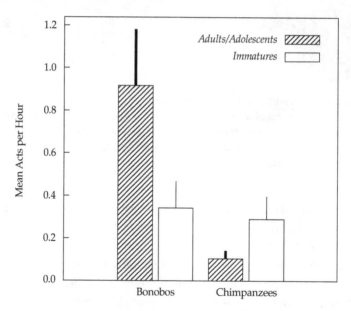

Figure 2.5 Mean (+ SEM) number of sociosexual acts initiated per hour per individual for the San Diego bonobos and an outdoor colony of chimpanzees at the Yerkes Primate Center's field station, separately for (a) adults and adolescents (hatched) and (b) younger individuals (white). The adult rate was significantly higher in the bonobo group than in the chimpanzee group despite the smaller number of available partners.

ners. Such conditions can be created in captivity, where I found sexual contacts to be far more variable and almost ten times more frequent among bonobos than among chimpanzees (see Figure 2.5). Confinement may well intensify sexual activity, but captive and field studies are in remarkable agreement about the form and function of sex in bonobo society. Inasmuch as the relatively high rate in bonobos cannot be attributed to the environment, it must reflect species-typical tendencies.

Incest Avoidance

Since incest may lead to inbreeding, and since inbreeding reduces the viability of offspring, animals generally avoid it (see Chapter 1). The bonobo's chief preventive measure is female migration. When adolescent females leave their mother and siblings to join a neighboring community, they go through a difficult transition that in the absence of significant benefits certainly would not be worth the risk. The trade-off is that the females get to mate with unrelated males, hence avoid inbreeding. Not that they do so knowingly: we assume that over the bonobo's history, females who migrated produced healthier offspring than females who did not.

When Chie Hashimoto and Takeshi Furuichi studied the sexual behavior of wild bonobos, they found that whereas males become more sexually active with maturation, females do not. On the contrary, the investigators speak of a "sexually inactive" state in young females in the period immediately preceding their departure from the natal group. Perhaps it keeps them from developing sexual relations with brothers and possible fathers. Females generally leave the group at the age of seven, when they develop their first little genital swellings. Equipped with this passport, they become "floaters" and visit neighboring communities before permanently settling in one. All of a sudden their sexuality flowers. They GG-rub with females and copulate with males encountered in strange forests. They have regular, almost continuous swellings, which grow in volume with every cycle until they reach full size at the age of perhaps ten. They can expect their first offspring by the age of about thirteen or fourteen years.

The situation is different for males, who remain in their natal group. Given that males cannot get pregnant, they do not risk much by having sex with relatives. It is the females in their group who stand to lose most from such contact. We assume, therefore, that mothers and sisters avoid sex with sons and brothers. These inhibitions are thought to come about through early familiarity, the basic mechanism assumed to underlie incest avoidance in a wide range of species, including our own. The principle is simple: individuals of the opposite sex with whom one has grown up since infancy fail to exert sexual attraction. If this process is disrupted—as when zoos raise young apes in a nursery—sex between relatives is not unusual. Normally, however, early familiarity characterizes the relations among close maternal relatives and keeps them from breeding.

Takayoshi Kano, the Japanese field-worker who has collected the most detailed information on the bonobo's natural history, concludes that incest avoidance is established early. After the age of two years, young males increasingly pursue sexual relations with females, but virtually never with their mothers. Kano recorded only five instances of sex in 137 mother-son pairs. There is a sharp decline in sexual involvement during male adolescence due to the tendency of dominant males to occupy the core of traveling parties, where the females are. Only when they enter adulthood and rise in rank, do males regain access to receptive females (see Table 2.1).

Table 2.1 Ten characteristics of bonobo social organization in the forest, developed from field studies at Wamba and Lomako, in the Democratic Republic of Congo

1. Like chimpanzees, bonobos live in male-philopatric fission-fusion societies.
2. Bonobos emphasize female bonding but show a potential for male bonding, whereas chimpanzees emphasize male bonding with a potential for female bonding.
3. Male kinship ties focus on the mother, rather than on brothers.
4. Food occurs in concentrations large enough that multiple bonobos can forage together, typically in mixed parties.
5. Bonobos are gregarious, forming large parties in the daytime and gathering together for the night.
6. The female hierarchy, based on age and residency, is relatively loose.
7. Males fiercely compete for rank, which is influenced by the rank of the mothers.
8. Females can monopolize prized foods: they tend to dominate males.
9. Despite some hostility between groups, peaceful mingling also occurs—an absolute difference between bonobos and chimpanzees.
10. Lethal aggression (infanticide, cannibalism, intergroup warfare) has thus far not been observed in bonobos.

Secondary Sisterhood

After a young female has left her natal group, she singles out one or two senior resident females in the neighboring group that she is joining and initiates sexual contact and grooming with these females to build a partnership. If the residents reciprocate, close associations are established and the younger female gradually becomes accepted and integrated. After having produced her first offspring, the young female's position becomes more stable and central until, as she grows older and achieves high status, the cycle is repeated, with young immigrants now seeking a comfortable relationship with her. Sex thus facilitates the migrant's entrance into the community of females, which is more close-knit for bonobos than for chimpanzees.

Observations by Amy Parish at the San Diego Zoo's Wild Animal Park demonstrate the distinct preference of bonobo females for each other's company. Eight different groupings were tried at the park, most of which included a single adult male, two adult females, and a couple of immatures. The adult partners varied: three different males and five different females figured in the rotations. Hundreds of records told Parish how much time individuals spent together, who approached whom, which partners groomed each other, and so on. The conclusion was that females favored the company of members of their own sex. Females sat together, groomed one another, and played together considerably more than with the male of their group. They actively pursued these contacts: females followed each other around seven times more often than they followed the male.

Preferential association among female bonobos is rather surprising, for it violates the rule that the sex that stays in the natal group develops the stronger within-sex bonds. For example, male bonding of chimpanzees follows naturally from males remaining in the community in which they are born. The same is true of female bonding in some Old World monkeys, such as macaques and baboons; males being the migratory sex in these species, females stay their entire lives together, forming complex kinship networks. Bonobos are unique in that the migratory sex seems to bond with same-sexed strangers later in life. Establishing an artificial sisterhood, female bonobos may be said to be secondarily bonded.[4]

Females occupy a central position in social life, so much so that Kano calls bonobo mothers the *core* of society. Males stay in their group and remain attached to the mother all their life, following her through the forest and depending on her for support against other males. During male power struggles, mothers often play a decisive role. This example from the Wamba field site shows how Koguma came to outrank Ude:

> The son of a powerful female, Aki, had begun to enter adulthood. This male, named Koguma, one day challenged the second-ranking male, Ude. Dragging a branch and screaming, Koguma charged straight at Ude, rushing narrowly past him. Ude leapt up and slapped Koguma in defense. The alpha male then intervened by mounting Ude, calming him down with rump-to-rump contact.
>
> After a while, Koguma charged again. Because Ude counterattacked, the two males ended up flying about between shrubs and bushes, exchanging violent blows. When Koguma launched another attack, his mother, with an infant clasped to her belly, came to assist him. Ude fled as Aki, with the loud calls of other females behind her, chased him off. Koguma did not let up: he attacked Ude 12 times in the span of nine minutes. Every time Ude rose to retaliate, Aki would go after him. Towards the end, Ude became silent and avoided Koguma. Eventually, he fled to the nearest tree whenever Ude branch-dragged in his direction.

As a result of such maternal interventions, the highest-ranking males in a bonobo community tend to be sons of the highest-ranking females. Male alliances are little developed, which allows females to exert considerable influence. A relatively young adult male can reach a top position, provided his mother is of high rank. On the other hand, males whose mothers are elderly or dead tend to drop in rank. Another report from the Wamba field site leaves no doubt that dominant males mate more often than others. Low-ranking males need to be secretive about their sexual exploits; they develop sneaky tactics to attract females. The link between male rank and sexual access means that by advancing their sons' careers, females may enhance their own reproductive success in terms of grandchildren.

This pattern is in strong contrast to chimpanzee males, who fight their own battles with each other, often relying on the support of other males. Both in captivity and in the wild, male coalitions are opportunistically established and broken in the pursuit of dominance—hence the title of my 1982 book, *Chimpanzee Politics*. Male chimpanzees regulate their own power plays and are fully dominant over females. Thus, an aspect of evolutionary scenarios that needed no adjustment when the chimpanzee replaced the baboon as the model of choice was male superiority. Baboon males are not only twice the size of females, they are equipped with formidable canine teeth, whereas females lack such weaponry. Sexual dimorphism is less in the chimpanzee, but males reign supreme: it is highly unusual for a grown, healthy male chimpanzee to be chased by any female. This constitutes perhaps the most conspicuous difference from the bonobo, whose females often dominate males.

At both major field sites in the Congo, as well as in captivity, females are able to claim food possessed by males (if necessary by force), and occasionally, as we have seen, they settle dominance disputes among their sons. Because bonobos show about the same degree of sexual dimorphism as chimpanzees this female power is rather surprising. When I first saw a female chase a male, I considered it an oddity. But the more captive bonobo colonies I studied, visited, or heard about, the more distinctly the pattern emerged as the rule rather than the exception.

That female dominance in a close relative is not accepted by all scientists is evident from a recent argument by Craig Stanford that we should look beyond food competition when considering the relation between the sexes. Stanford's main source of information on bonobo relations was a brief report by scientists who, having worked at Lomako without recognizing individual bonobos, questioned the generality of female dominance. Instances in which females did supplant or chase males were dismissed as male deference, even chivalry! However, Barbara Fruth, another investigator at the same site, reached the following conclusion after a much more intensive study that did involve individual recognition: "Females show a high degree of association, form coalitions, and dominate the society."[5]

When a male chimpanzee chases a female away from food, we generally do not hesitate to attribute this activity to his dominance. We have done so for every species on the planet, so it is unclear why we should suddenly adopt a different standard for bonobos. Kano has forcefully rejected the argument:

> Priority of access to food is an important function of dominance. Since most dominance interactions, and virtually all agonistic episodes between adult females and males occur in feeding contexts, I find much less meaning in dominance occurring in the non-feeding context. Moreover, there is no difference between feeding and non-feeding dominance relationships among the bonobos of Wamba. For example, approaches of dominant females often give rise to submissive reactions by grooming males such as grinning, bending away, etc.

Mothers against Infanticide

Given the amount of sex that goes on in bonobo society, and given that female receptivity stretches well beyond the brief period of ovulation, the male who knows his offspring must be a genius. He would need to calculate, perhaps from observed menstruations and births, which females might be fertile and favor them as sex partners—or at least keep track of when he had intercourse with them. Male bonobos are obviously not keeping records; they are merely attracted to large, pink genital swellings. That even pregnant females sport these "signs of fertility" makes no difference to them. Consequently, the bonobo male has no idea which copulations may result in conception and which may not. Almost any young ape growing up in the group could be his, but it could also have been sired by almost any other male, including those in neighboring territories (owing to copulations during intergroup mingling). In short, bonobo society is marked by an extremely low confidence in paternity.

If one had to design a social system in which fatherhood remained obscure, one could hardly do a better job than Mother Nature did with bonobo society. Based on the ideas of Sarah Hrdy about concealed ovulation, several scientists have begun to speculate that this may in fact be the whole purpose. Females may sport almost continuous swellings to lure males into frequent sex. No conscious intent is implied, only a systematic misrepresentation of fertility in the service of female reproduction. But what could be wrong with males knowing which offspring they have sired?

Sometimes chimpanzee males, as well as the males of quite a few other animal species, kill newborns (see Chapters 1 and 3). They do not do so often, but they do so frequently enough to pose a grave problem for females. This male behavior produces a setback in reproduction and a loss of investment in gestation and lactation. What gain do the infanticidal males obtain? Inspired by work on Indian langurs, Hrdy's theory proposes that this is the way a new male who takes over a group forces females to restart their reproductive cycles. By eliminating their newborns, the male makes sure that the females will soon become available for sex and conception. Instead of waiting years for them to resume cycling, the infanticidal male improves his chances at reproduction almost immediately at the expense of the rival ousted from the group when he took over.

This proposed evolutionary scheme only works if a male can be relatively sure that he himself did not father the infant he eliminates. Killing

from harassing a female and killing her offspring. Thus, gibbon monogamy may have arisen from a male tendency to accompany females with whom they had mated, so as to repel infanticidal males—to protect their genetic investment. The same argument can be applied to the human nuclear family, in considerable contrast to the views of Lovejoy, Fisher, and others, which stress male care giving and food provisioning. Both industrialized and preliterate societies manifest evidence that children whose fathers have left or died are disproportionately victims of abuse and infanticide. Stable pair bonds seem to protect the young.

Among our hominid ancestors, a simple security arrangement between the sexes would have been easy to expand upon. For example, the father could have helped his companion in locating fruit trees, capturing prey, or transporting juveniles. He himself may have benefited from her talent for precision tool use, and the gathering of nuts and berries. The female, in turn, may have begun prolonging her sexual receptivity so as to keep her protector from abandoning her. The more both parties became committed to this arrangement, the higher the stakes. From an evolutionary perspective, investment in someone else's progeny is a waste of energy; therefore, males may have tightened control over their mate's reproduction in direct proportion to their assistance to her. (Wrangham in Chapter 5, discusses a different, but also mutually beneficial, arrangement between the sexes.)

If these arguments are valid, the societies of humans and great apes part company when it comes to attitude toward offspring. The chimpanzee male has a tendency to target infants who are *not* his so as to increase his own reproduction by eliminating them. The female counterstrategy is to confuse the issue of paternity: if all males are potential fathers, none of them has a reason to harm newborns. In the bonobo, females apply this strategy with perhaps even greater sophistication and combine it with collective dominance over males, a double strategy that may have been successful in eliminating infanticide altogether. In our own species, in contrast, males do everything to ensure that they father their mate's offspring, and females do everything to ensure male protection: the nuclear family rests on paternity certainty.

Whence the Difference?

The initial stimulus in the bonobo's social evolution probably occurred when females began to have more frequent and longer-lasting genital swellings. Together with a general "sexualization" of the species, this trend reduced competition among males, obscured paternity, and promoted sociosexual relations in all partner combinations, particularly among the females. The end result was that females formed a secondary sisterhood, gained the upper hand in society, and freed themselves from the curse of infanticide.

This script might never have worked without large enough food concentrations to permit the association of multiple males and females. How could female solidarity ever have been achieved, and how could it effectively operate, if bonobo communities in their quest for food were forced to split into smaller parties? This situation holds for some of the best-known chimpanzee populations, and it may explain why the species has not gone down the same evolutionary path as the bonobo. Fieldwork supports the assumption that the distribution and abundance of food is crucial to the contrasting life styles of the two *Pan* species.

Three anthropologists—Richard Malenky, Frances White, and Richard Wrangham—compared the feeding ecology of the Lomako bonobos with that of chimpanzees. They concluded that bonobos have access to larger fruiting trees that allow more individuals to feed together, and that the bonobos consume larger quantities of terrestrial herbaceous vegetation, a common food source in their range. They bite open the tough fibrous sheath and chew the soft pith of canelike herbs. The investigators believe that the social structure of bonobos reflects the presence of these predictable, protein-rich food sources on which multiple females can feed without conflict.

Since herbaceous foods are a staple of another close relative, the gorilla, Wrangham has speculated that the total absence of gorillas in the bonobo range may have opened an ecological niche that is unavailable to chimpanzees (who are sympatric with gorillas in large parts of their range). Released from feeding competition with gorillas, the ancestors of bonobos were able to turn to the abundant herbaceous foods on the forest floor. The protobonobos could forage together in large parties, because—unlike chimpanzees—they were not forced into small parties by competition over scarce fruiting trees.

The initial stages of this evolutionary scenario remain the most obscure. The so-called sexualization of the bonobo—meaning that sexual behavior began to permeate all aspects of social life—most likely started with relations between the sexes. After all, the original function of sex is reproduction, which implies adult heterosexual relationships. Perhaps both sexes gained from each other's company and from tolerant and friendly relations promoted by frequent sex. Since female mammals almost never dominate adult males of their species (notable exceptions are the lemurs of Madagascar and spotted hyenas), it is safe to assume that bonobos started out under male dominance. In this regard, therefore, the common *Pan* ancestor may have been more chimpanzeelike than bonobolike. Sex-for-food transactions may have helped females gain access to food controlled by high-ranking males. Consequently, females gained from the extension of sexual receptivity.[7]

Conceivably, this early role of sex as the cement of society spread from the heterosexual domain to other domains. Sexual activity in all combinations became a way of tying males and females together in ever-larger aggregations. The question of why bonobos and not chimpanzees sexualized their social relations is hard to answer, however, without reference to infanticide. Sexual behavior is by no means required for bonding and tolerance. Chimpanzees, for example, employ nonsexual means that seem as effective as the genital contacts of bonobos. Chimpanzees reconcile with a kiss, and they share food after loud vocalizations and body contact in which they hug and pat each other on the back. The reliance of bonobos on sexual mechanisms may derive from the added benefit of obscured paternity, especially in a species in which males and females travel and forage together.

That male bonobos do not band together to put a halt to female ambition is perhaps because the payoffs of being dominant are not substantial enough to achieve effective male-male cooperation. In all animals, males compete over mating rights, which makes alliance formation a sensitive process. It hinges on what each party to the alliance, not just the eventual winner, gets from it. Few animals are capable of striking mutually profitable deals. Chimpanzee males are a notable exception: dominant males selectively share food and sexual privileges with male allies to whom they owe high status.

There is much less reason for male cooperation in a species such as the bonobo, in which sexually receptive females are relatively plentiful. What would be the point, for example, of two or more males guarding a particular female if other receptive females are freely available to their competitors? While it may pay for males individually to compete over females, the advantage of monopolizing a female may never reach the threshold at which alliance formation becomes an advantageous strategy.

Apart from the lack of a sound basis for male cooperation is the rising influence of the mother. The higher the status of females became, the more mothers could do to help sons in their quest for high rank. When maternal support became as effective for males as support from other males, it further undermined whatever tendencies for male cooperation may have existed. Reliance on a partner, such as the mother, with a stable preference became increasingly advantageous. Eventually the nucleus of traveling parties shifted from association between members of the opposite sex to association between mother and son as well as the powerful alliance among senior resident females.

The scenario delineated above (Figure 2.6) must be regarded as tentative. Even if not accepted in its entirety, it still has plausible components:

1. Extended female receptivity dilutes competition among males. The presence of so many attractive females, and the impossibility of pinpointing their fertile days, makes it less worthwhile for males to risk injury over a specific mating.
2. Given that male alliances in other primates are mostly instruments to keep competitors away from a highly contested female, or to attain high status that translates into access to females, the basis for such cooperation is eliminated if multiple females are sexually attractive at once.

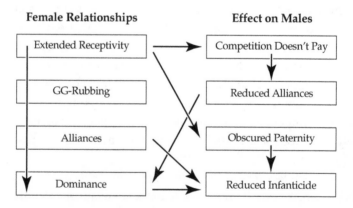

Figure 2.6 A hypothetical scheme connecting evolutionary developments in bonobo social behavior. The female bonobo's extended sexual receptivity reduces male competition, which paradoxically may result in a lower degree of male cooperation. Female sexual relations also appear to aid bonding, which translates into effective female alliances. Finally, the frequent sex in bonobo society reduces paternity certainty, resulting in a reduced likelihood of infanticide by males. (First published in de Waal, 1997a. Used with permission of the author.)

3. An every-male-for-himself system paves the way for a collective female power takeover.
4. Sociosexual behavior and bonding among females translates into alliances that allow them to monopolize food and protect their offspring against infanticidal males.
5. Extended receptivity and frequent sex confuse paternity to such a degree that infanticide becomes counterproductive: males have trouble exempting progeny.

Human Evolution

However plausible this scheme may sound, it is not without problems. It assumes that bonobos descended from chimpanzeelike ancestors, whereas other scientists believe that bonobos represent the ancestral form. This issue is not easily resolved. It is of course entirely possible—even likely— that the common ancestor of bonobos and chimpanzees was both chimpanzeelike, as by male dominance, *and* bonobolike, as by the prominent social role of sex. After the two species split, the first set of characteristics may have been amplified in the chimpanzee's evolution, and the second set in the bonobo's.

The scenario also raises the question of why, if ecology shapes social organization, bonobos did not adopt a more gorillalike social organization in which a number of females are led by a male who competes with other males over their "possession." The bonobo seems to have followed almost the opposite evolutionary path: large domineering males are nowhere in sight!

Perhaps the chances for a gorillalike system were slim from the beginning, as abundant foods in large patches allowed females to associate, bond, and eventually form alliances that held male ambition in check. If so, a slight initial advantage in the battle of the sexes may have allowed bonobo females to achieve a totally different social organization in which not males but the females themselves were in control.

Did our social system derive from one similar to the bonobo's? With a bonobolike ancestor as starting point, human evolution would have required dramatic changes, such as the loss of female genital swellings while retaining prolonged receptivity, an overturn of female dominance, the evolution of male cooperation in hunting and perhaps warfare, and establishment of the nuclear family. How and why these changes might have taken place is ambiguous; it seems more economical to assume that the common ancestor of apes and humans had no swellings at all and was characterized by both male dominance and male-male cooperation. With regard to the latter aspect, clear parallels exist in the chimpanzee's social system. However, when it comes to female bonding and the nonreproductive use of sex, bonobos seem the better models.

For the moment, the safest assumption is that all three species—humans, chimpanzees, and bonobos—are specialized, that is, that they evolved considerably after their descent from the last common ancestor. Unresolved is the puzzle of the common ancestor's social life: no extant species can be adopted as model. However, in the bonobo we have an additional key to unlock the mystery: this species may have retained different traits of the ancestor than did both other species. The bonobo shows an unparalleled social organization that should give pause to anyone claiming the universality of certain traits in our ancestry. The vast difference in two close relatives such as the bonobo and the chimpanzee hints at a flexibility in our lineage that many of us had not held possible.

It also contains a warning for paleontologists who are reconstructing social life from fossilized remnants of long-extinct species. On the sole basis of a few bones and skulls, no one would have dared to propose the dramatic behavioral differences recognized today between the bonobo and the chimpanzee. True, the smaller canine teeth of the bonobo hint at reduced male competition, but certainly neither the central position of females, nor the extensive use of sex, nor the reduced cooperation among males seems to follow logically from bonobo anatomy.

Most important, we should keep in mind that the bonobo and the chimpanzee are equidistant from us. Most biologists consider the general principles of adaptation and natural selection more important than the evolution of a particular species. It is true that the bonobo, by virtue of its close relation to us, is a critical piece in the puzzle of human evolution; but it is really the entire puzzle that science seeks to solve. The most successful reconstruction of our past will be based on a broad, triangular comparison of chimpanzees, bonobos, and ourselves within this larger evolutionary context.

3

Beyond the Apes:
Reasons to Consider the
Entire Primate Order

Karen B. Strier

What is special, or not so special, about the hominoids? Comparisons between humans and the anthropoid apes are highly informative, but we cannot make sense of this small taxonomic group's characteristics without taking a broader comparative perspective that encompasses the two hundred other extant primate species.

N O O N E D I S P U T E S the special place that chimpanzees and bonobos hold in comparative models of human social evolution. They are the only two living species of the genus *Pan*—and they share more than 98 percent of our DNA. By some estimates, all that distinguishes us from these apes is encoded in some 50 or so genes. These differences accumulated during the 5 million to 7 million years when the lineages leading to modern humans and *Pan* diverged from one another. This shared common ancestry makes the behavioral heritage of chimpanzees and bonobos an undeniable part of our own.

With such close relatives to turn to for clues, it is tempting to ignore the other two hundred species of more distantly related primates and focus exclusively on the social similarities between humans and these apes. For example, consider the simplistic characterizations of chimpanzee societies, which like many traditional human societies are based on strong kin bonds among males who cooperate in territorial defense, and of bonobo social relationships, which like those of many humans are reinforced by sexual interactions that extend outside the reproductive realm into nearly every aspect of daily life. Both the capacity of male kin to cooperate in territorial defense and the use of nonreproductive sex are thought to have played influential roles in shaping the social possibilities of human ancestors, so understanding how these behavioral tendencies influence chimpanzee and bonobo societies can provide insights into the origins of our own. But if we assume that the social similarities between humans and these apes are an inevitable result of our common ancestry, what are we to make of the compelling evidence that many other male primates also cooperate with their kin, and that engaging in sex for social as well as reproductive ends is prevalent across the primate order?

Common ancestry cannot account for cases of behavioral convergence among distantly related species, nor can it explain the divergent social tendencies of closely related species such as chimpanzees and bonobos, which have persisted for nearly 2 million years since their respective ancestors embarked on separate evolutionary paths to develop the behavioral peculiarities that distinguish them today. To make sense of the behavioral differences between chimpanzees and bonobos, as well as their similarities to other primates, including ourselves, requires adopting a broader comparative perspective that extends beyond the apes.

The idea of looking at the social behavior of other primates for insights into human behavioral evolution is not new. In the late 1950s, the American physical anthropologist Sherwood Washburn and one of his students, Irven DeVore, set out to study savanna-dwelling baboons with precisely this goal in mind (see Figure 3.1). Our own bipedal ancestors were thought to have evolved under similar ecological conditions, and might therefore have made social adjustments similar to those made by baboons.

Figure 3.1 East African baboons in their savanna habitat. I was fortunate to have the opportunity to watch a well-known baboon troop, whose members had been fully habituated to humans and extensively studied by my predecessors. As the baboons climbed down from their sleeping trees each morning and set out on their foraging forays across the East African savanna, I could see why they had for so long held the attention of anthropologists interested in human social evolution. (Photo by K. B. Strier.)

Like early hominids, baboons encounter an entirely different set of challenges than arboreal primates each time they leave the safety of the trees. On the ground, they are vulnerable to attacks from large terrestrial predators, so they need to live in large cohesive troops with many pairs of watchful eyes and alert ears. Large troops could also be advantageous in tracking down scarce food and water sources and defending them from competing baboon troops, while strict social hierarchies among troop members (in particular among females) reduce the constant need for contests over prized feeding and drinking sites.

Out on the open savanna, baboons can see one another most of the time. Thus, it is difficult for them to conceal their activities behind dense foliage in the way arboreal primates can. Their increased visibility also puts a premium on sophisticated social skills that involve appeasement instead of avoidance, and may contribute to the high rates of social interaction that characterize baboons and most other open-country primates who also live in cohesive groups. Indeed, the close associations between female baboons give them many more opportunities to interact than the more flexible, ephemeral associations formed by their chimpanzee cousins.

To pioneering baboon watchers like Washburn and DeVore, there was little doubt that males, roughly twice the size of females and with disproportionately larger canines, were in command of their troops. Through bullying and harassment, domineering males seeking reproductive opportunities could presumably have their way with females. On the surface, there seemed to be little a female baboon could do to avoid attracting unwanted sexual attention or succumbing to sex with the most powerful males. And because male sexual interest picked up markedly whenever females displayed the bright pink genital swellings that signal the most fertile times in their cycles, it was only logical to assume that sex and reproduction were closely coupled in baboons, and by inference in most other primates.

Both models of hominid ecology and our understanding of baboons have changed over the years. Early depictions of heroic hominids stepping out of the forest onto a hostile savanna have given way to scenarios of more benevolent, transitional woodland zones that hominids opportunistically exploited. Similarly, early emphasis on males and their capacity for aggression and troop defense as the core features of baboon society has been replaced with discoveries about the relatively transient troop memberships that males maintain compared to the lifelong stability and social influence of extended female matrilines.

In fact, the savanna baboon pattern in which females remain as lifelong residents of their natal groups and males disperse to join other groups was first discovered by Japanese primatologists studying Japanese macaques, another member of the Old World monkey family Cercopithecinae, to which baboons also belong. Male-biased dispersal with female residency has since been confirmed in most of the other cercopithecine species that have been studied. The result has been a long-standing perception that social groups composed of female kin are widespread across nearly all primates.

In the context of the underlying cercopithecine and baboon biases, it is not so surprising that discoveries of different behavior in the apes seemed exceptional. The strong kin bonds among male chimpanzees and the sexual side of the bonobo's egalitarian society were, and still are, completely anomalous by baboon standards. If apelike behavioral oddities had first been described in any other primate, they would probably have been subjected to intense initial scrutiny, then cast aside as no more than interesting deviations from the cercopithecine primate rule. Indeed, by the late 1960s and early 1970s, other examples of such deviant primates were known. But finding the same apparent anomalies in our closest relatives was another matter altogether, and more difficult to dismiss. After all, we know that we are different from other primates, so why should our closest nonhuman relatives not differ as well?

The problem with thinking in such anthropocentric, or human-centered, terms is that it leads to setting chimpanzees and bonobos, along with humans, apart from other primates on the basis of ancestry alone. In contrast, where we consider the behavioral diversity of primates other than apes, the social tendencies that humans share with chimpanzees or bonobos are situated along a continuum defined by common responses to common ecological challenges. Our history as hominoids predisposes us to act like apes, but underlying these predispositions is an even longer past as socially adaptable primates.[1]

Behavioral Homology and Adaptation

Species can end up possessing similar traits through two alternative routes. One route involves the mutual inheritance of traits from a common ancestor and results in what are known as homologies. The other route can be traced to similar adaptive solutions to ecological challenges that arise independent of the ancestral relationships between species. Together, homologous traits and derived adaptations make up a primate's behavioral repertoire, which may include the flexibility to respond in different ways to fluctuating or changing ecological conditions. Of course, the abilities of primates, like all other organisms, to adapt to their environments are constrained by the traits they already possess. Just as the lack of anatomical precursors for flight prevents primates—but not birds, bats, or many insects—from flying, the range of behavioral options available to primates is also dictated in part by their particular evolutionary histories. Because ecological pressures and evolutionary history interact in their effects on behavior, the only way to distinguish between behavioral homologies and shared, derived adaptations is through careful, broad-based comparisons.

Out-group Comparisons

Distinguishing behavior patterns that are inherited through ancestry from those that are sensitive to local ecological conditions requires comparing closely related species, whose common ancestry aligns them into clades, with more distantly related species, known as "out-groups" when used for this purpose. Technically speaking, humans are the most appropriate out-group for interpreting behavioral differences between chimpanzees and bonobos if we consider ourselves to be the next-closest relatives of these sister taxa. However, if (as some of the molecular data suggest) gorillas and humans are equally related to the common *Pan* ancestor, then the African hominoid clade should include four living species, and we should consider the Asian apes—orangutans and their smaller gibbon cousins—as our comparative out-groups.

Different pictures of behavioral evolution can emerge depending on how ancestral relationships are defined, because the clustering of species into clades affects the choice of comparative out-groups. For example, the societies of chimpanzees, bonobos, and human foragers are characterized by female-biased dispersal and lifelong group residency of natal males, or male philopatry. In gorillas, orangutans, and gibbons, by contrast, it is more common for both sexes to disperse. If we take dispersal by both sexes to be the more primitive, or ancestral, hominoid pattern, then the occurrence of extended male kin groups with female-biased dispersal emerges as a novel, or derived, condition. Whether male philopatry was shared by the last common ancestor of humans and *Pan*, or independently derived in both humans and *Pan* after their lineage split, depends on where gorillas figure into the phylogenetic framework (see Figure 3.2). If gorillas are appropriate out-groups to a human-chimpanzee-bonobo clade, then male philopatry in the clade could be a behavioral homology, as has often been assumed. However, if gorillas actually belong alongside us in an African hominoid clade, then the case for homology is less clear-cut.

Old World cercopithecine monkeys, like the well-known savanna baboons and various species of macaques, have traditionally been used as the out-groups for identifying how the social tendencies of hominoids differ from those of other primates. No hominoids are known to live in cercopithecine-like extended female matrilines with male-biased dispersal, so from a cercopithecine perspective, the ancestral hominoid pattern in which females as well as males disperse appears to be derived.

Phyllis Lee has theorized that the ability of a species with male- or female-biased dispersal to evolve into one in which dispersal is biased in favor of the opposite sex should be constrained by the deleterious genetic consequences of inbreeding. These would arise if close male and female relatives were restricted to mating with members of their natal groups during the inevitable transitional phase of such a drastic behavioral shift.

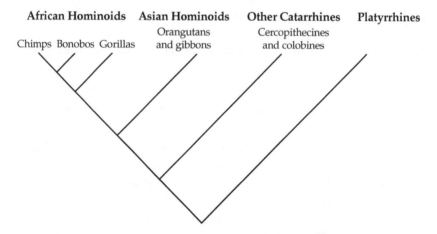

Figure 3.2 Anthropoid out-groups for behavioral comparisons. Some authors emphasize genetic evidence supporting a three-way split between the two species of *Pan* (chimpanzees and bonobos), gorillas, and humans. Asian hominoids are appropriate out-groups for comparison among African apes, which together with the cercopithecine and colobine monkeys constitute the catarrhines. Platyrrhines, which include all of the New World monkeys, complete the anthropoid suborder. Prosimians, which include lemurs, bushbabies, lorises, and tarsiers, are the other suborder of Primates.

However, if both sexes disperse, sex biases in primate dispersal patterns can arise without genetic costs.

The fact that male-biased dispersal characterizes nearly all contemporary cercopithecine species implies that it is a homologous behavior pattern, with evolutionary antiquity in this taxonomic group. The assumption that male-biased dispersal also represents the ancestral primate condition from which hominoid dispersal patterns derived is not tenable when the dispersal patterns of primates other than cercopithecines are considered. In fact, we now know that cercopithecines are unusual primates in many respects.[2] Among the colobines, which constitute the other major group of Old World monkeys, dispersal by both sexes occurs, as well as male-biased and female-biased dispersal. Among the New World monkeys, it is only among capuchins that male-biased dispersal and female residency are consistently present. Other New World monkeys exhibit more variable patterns in which either both sexes or only females disperse. In fact, if we accept that female or male philopatry is only likely to arise from an ancestral condition in which both sexes dispersed, then the dispersal pattern of ancestral hominoids is a primitive one, and the emergence of male philopatry in *Pan* and humans is no more special than its emergence in some species of Old World colobines and many species of New World monkeys (see Figure 3.3).

Extended out-group comparisons, which include the colobines and New World monkeys, shed a very different light on *Homo-Pan* similarities than do more restricted comparisons with cercopithecines as the out-groups to the hominoids. Sex biases in primate dispersal patterns are fundamental because they determine whether social groups are composed of male or female kin, and therefore set the stage for different kinds of social options. Specifically, biological kin have a stake in one another's reproductive futures because they share some portion of their genes. This may be why young male gibbons are tolerated on adjacent territories held by male relatives who are otherwise hostile toward unrelated male intruders, and why silverback male mountain gorillas sometimes permit their adult sons to stay on as subordinates in their natal groups. Together, these father-son gibbon or gorilla teams can put up more powerful resistance to male outsiders than either partner could achieve on his own.

Usually, kin join forces against unrelated competitors for such purposes as the collective territorial defense engaged in by philopatric male chimpanzees. However, the absence of similar displays of solidarity among male bonobos is evidence that cooperation is not an inevitable outcome of philopatry among males. Perhaps even more remarkable, close kinship is not always a condition for cooperation, at least with regard to chimpanzee coalitions or multimale mountain gorilla groups.

Comparing Behavior

Chimpanzees and bonobos are not exceptional in being closely related species with fundamentally different social strategies. In fact, behavioral diversity, both between different species and among populations of the same species, is probably more widespread across primates than any other mammalian order. Some of this diversity can be explained by the ways in which local ecological conditions, such as predator pressures and the seasonal and spatial distribution of food resources, shape behavioral responses. For instance, whether primates live in cohesive or fluid groups can usually be predicted by measurable ecological and demographic variables that differ from one population to the next. Even contrasts between the male-dominated hierarchies of chimpanzees, on the one hand, and the

STRENGTH OF MALE KIN BONDS

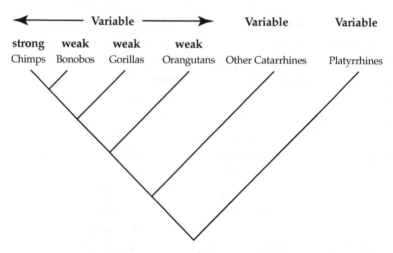

Figure 3.3 The strength of philopatric male kin bonds. Variability in the type and strength of these bonds distinguishes chimpanzees from bonobos and other hominoids, but comparable patterns of variable male kin bonds are found in other primate species. Variability in male kin bonding in some neotropical genera is comparable to the differences between chimpanzees and bonobos.

more egalitarian relationships of bonobos, on the other hand, come down to the degree to which individuals or groups can monopolize resources, such as food and mates, that are important to reproduction.

Not all aspects of primate social behavior are as facile in the face of local ecological conditions as grouping patterns or the types of social relationships that individuals maintain with one another. For example, dispersal tendencies seem to be more resistant to change than other social characteristics of a species. As a result, closely related species are likely to share similar dispersal patterns despite the powerful influence of ecology on other aspects of their social lives.[3] Yet similar dispersal patterns do not always translate into similar social relationships, as the differences between philopatric male chimpanzees and bonobos so clearly show.

The Muriquis of Brazil

Deciphering patterns of primate kinship and social relationships often comes down to time and circumstance. Like other primatologists studying poorly known primates, I found few established guidelines when I began to investigate New World monkeys. The target of my efforts was the muriqui, one of the most elusive and endangered of the New World monkeys, found only in the remnants of Brazil's Atlantic Forest (see Figure 3.4). Muriquis were ideal test subjects for evaluating hypotheses about the very different social systems found in the multimale, multifemale societies of primates like chimpanzees, bonobos, and baboons. I knew how baboonlike primates were supposed to behave, and also how chimpanzees and bonobos differed from them. I also knew that the spider monkey, one of the muriqui's closest relatives, was known to resemble the chimpanzee with its dietary preference for ripe fruits, fluid female grouping patterns, and male cooperative coalitions.

Where muriquis would fit into this picture was anybody's guess. They were known to travel by means of suspensory locomotion, which meant they could move quickly from one patch of ripe fruits to another as spider monkeys did, and might therefore maintain similarly fluid grouping patterns to avoid competing for food. Yet muriquis were also known to have teeth and guts equipped for processing more abundantly occurring foods such as leaves. In other primates, including bonobos, access to leaves and herbaceous vegetation allows for more cohesive grouping patterns than those found in specialized fruit-eaters such as spider monkeys and chimpanzees. Moreover, male and female muriquis were known to be physi-

Figure 3.4 Muriqui monkeys from the Brazilian Atlantic Forest. Despite their evolutionary and ecological distance from the apes, muriquis nonetheless exhibit striking similarities in various aspects of their social behavior. Most notable are the highly affiliative, egalitarian relationships that philopatric males maintain with one another and with females in their groups. (Photo courtesy of Claudio P. Nogueira.)

cally more similar to one another in their body and canine size than both chimpanzees (where males are bigger and have disproportionately larger canines than females) and spider monkeys (in which males have larger canines). Such sexual dimorphism in these and most other nonmonogamous primates, especially in canine size, is usually linked to high levels of male-male aggressive competition and male dominance over females. The evidence for little if any canine dimorphism in muriquis was reason to suspect that the social dynamics governing their society would be different as well.

My first year of study involved documenting their diet and the social interactions between members of different age and sex classes who could be individually recognized from their natural hair coloration and facial markings. Although they ranged widely through the forest to find their favorite fruits, muriquis devoted most of their feeding time to leaves. Group members spent the majority of their time together and treated one another with tremendous tolerance. They almost always avoided direct contests over food, and males were uncommonly courteous toward one another over access to mates. Although aggressive interactions were rare among group members, the males showed strong solidarity when they took a stance in encounters against males from neighboring muriqui groups.

Except for the close social bonds that muriqui mothers, like all primates, maintain with their dependent offspring, there were no shortcuts to figuring out other kinship relationships within the group. It took years of continuous monitoring to compile the maternal genealogies on which most of our knowledge of kinship in primate groups is based. In general, primates have long interbirth intervals and mature slowly compared to most other mammals of similar body size, so understanding their kinship systems without genetic data can come only as fast as the animals mature and reproduce. It took three years to identify maternal siblings based on what my students and I discovered to be the length of muriqui birth intervals, and about seven years before I could spot a mother embracing an adult male I knew to be her son.

When a juvenile, an adolescent, or even an adult disappears from a study group from one day to the next, either predation or dispersal can be plausible causes. But, during that first year of research, none of the muriquis disappeared. Two unfamiliar adolescent females turned up in succession, but it was years before I could confirm my initial suspicion that in muriquis, sons stay where they are born and daughters typically disperse.

The discovery of male philopatry and female dispersal in muriquis made sense in light of their common ancestry with spider monkeys, who follow a similar pattern. Like many other New World monkeys, both have slower rates of maturation and longer interbirth intervals than Old World monkeys with similar body sizes. In fact, in their life histories and correspondingly slow reproductive rates, as well as in their female-biased dispersal patterns, these New World monkeys resemble the apes.

The discovery of male philopatry in muriquis also helped to make sense out of some of the other social patterns that they display—along with spider monkeys, chimpanzees, and bonobos. For instance, kinship among males is consistent with their solidarity against males from other groups, and is probably one of the factors contributing to their tolerance of one another's sexual liaisons. After all, a male who misses a chance to fertilize a female has less to lose in terms of his genetic fitness when his competitor is a close relative rather than a distant relative or altogether unrelated.

Kinship can temper the degree to which males compete for mates, but it does not prevent either chimpanzee or spider-monkey males from fighting for high rank. This goal translates, at least to some extent, into greater mating and presumably reproductive success. The near-absence of overt aggression displayed by muriqui males—either in front of fertile females or at other times—carries kin selection and the tolerance it favors to an unprecedented extreme.

Clearly, male primates take advantage of the presence of their kin in markedly different ways. These differences are expressed in the maintenance of affiliative relationships, which can be measured by the amount of time males spend together and by the frequency of amicable interactions, like grooming and embracing, in which males engage with one another versus with other group members.

Such affiliative displays are important to secure solidarity whenever males rely on their male or female kin for support in internal power struggles, as well as in intergroup conflicts. The extraordinarily peaceful relationships that male muriquis maintain seem to obviate the need to cultivate special allies within their groups. In other primates, whether males turn to one another or to females for social support tends to coincide with the social positions of males and females in their societies. Thus, male chimpanzees may be fickle about choosing sides in contests among themselves, but they put their individual animosities aside in collective defense of their communal territories. Similarly, the recent finding by the Japanese primatologist Takeshi Furuichi that bonobo mothers play decisive roles in settling dominance disputes among their fully grown sons is consistent with the prominence bonobo females play whenever the infrequent inter-group conflicts erupt.

Two additional factors regarding females figure into the social equation of male behavior, and like the social contrasts between chimpanzees and bonobos, or between muriquis and spider monkeys, neither has much to do with phylogeny or the effects of male kinship.

Influence of Female Grouping Patterns

For starters, female chimpanzees and spider monkeys, and to some degree muriquis, are less likely than female bonobos to be in the same place at the same time. Female primates, with strong dietary preferences for patchy foods like fruits, tend to adjust their grouping patterns in response to the size and temporal availability of their favorite food patches. Fluid associations of females are harder for males to track than females who manage to maintain more cohesive groups by incorporating into their diets abundant foods such as the herbaceous vegetation that bonobos eat or the leaves on which muriquis rely when fruits are seasonally scarce.

Monitoring scattered females puts a premium on male cooperation, which is necessary for their collective territorial defense. Moreover, when females go off on their own in search of food, they are less likely to be available to males as allies. Interestingly, as Frans de Waal has shown, where female chimpanzees are housed alongside males in captive colonies instead of being free to forage on their own as they do in the wild, females enter into and strongly influence male social dynamics.

In terms of their cooperation in intergroup conflicts, chimpanzees resemble the more distantly related but ecologically more similar spider monkeys much more than they resemble bonobos. Although common ancestry provides clues to the patterns of male philopatry, ecology evidently is a better predictor of the strength and kinds of social bonds among these philopatric males.

Female Social Influence

A second way in which females affect male social dynamics is more direct. Among muriquis, for example, we have found that some males become more popular with age, as indicated by the degree to which younger males seek out their elders for company during long resting bouts or for quick, reassuring embraces during times of tension. But age has little bearing on a male's popularity with sexually active females, whose equal body and canine size seems to give them equal status, and with it the freedom to chose their own mates.

Muriqui females do not copulate year-round. Usually it is only at the onset of the annual rainy season, in late September, when those females unencumbered by nursing infants become increasingly attractive to males. First the frequency with which females present their genitalia to males for inspection begins to rise. Then copulations with compliant females become common events. By May, the start of the dry season, most sexual activities have long since ceased, and the females who have conceived begin to give birth.

A striking feature of muriqui mating patterns is that males make no overt moves to interfere with one another's sexual activities and do not try to harass females into copulating with them. Instead, males wait patiently until a receptive female favors them with a chance to mate.

The ability of female muriquis to express their mating preferences seems to override whatever reproductive rewards high rank confers in other primate societies, where dominant males can sometimes pressure reluctant females into cooperating as consorts. Fighting among males is futile in the face of unmonopolizable mates, so muriqui males rely for fertilizations on female favoritism and perhaps on more subtle forms of competition.

Male muriquis have disproportionately large testes relative to their body size. Copulations usually lead them to produce copious ejaculate, which hardens almost instantly into a solid plug that is clearly visible to observers and presumably to other muriquis. We cannot be sure that the muriquis' large testes correspond to high sperm counts or rapid rates of sperm production; conducting the required experiments to check these details is out of the question, given the muriquis' endangered status. Nor do we know what function their postcopulatory plugs serve, especially considering that it is not uncommon for a female (or, with her permission, another male) to remove a previous suitor's plug and eat it.

What we do know is that, in muriquis and other primates, female freedom from the threat of male aggression not only inhibits males from behaving like bullies but may even encourage cooperation among male kin. For example, among Peruvian squirrel monkeys and most species of lemurs, females are socially dominant over males. Although males in these species typically disperse from their natal groups, they typically do so in pairs. Peruvian squirrel monkey brothers form long-term partnerships to fight their way into new groups, where they collaborate in their courtship of females. Cohorts of dispersing male ring-tailed lemurs sometimes stay together for years, moving from one female group to another in search of favorable sex ratios and reproductive opportunities. Thus, despite the absence of male philopatry in these species, the social influence of females can exert a powerful force that makes it advantageous for dispersing males to keep close track of their kin. The males thereby have access to related allies who can smooth their interactions with socially dominant females as well as aid them in contests against other males.[4]

Comparative Insights into the Social Aspects of Sex

The freedom that bonobo females enjoy with respect to males has been attributed to their active sex lives, which cement social bonds with other females as well as with the males they chose as mates (see Chapter 2). Like their chimpanzee sisters and baboon cousins, bonobo females have fleshy patches of genital skin that swell and change color in ways that males find attractive. In chimpanzees and baboons, these swellings are honest advertisements of female fertility; consequently, male chimpanzees and baboons time their sexual advances toward females—and take their strongest competitive stances with one another—to coincide with the odds of ovulation and the probability of conception. Between their swellings, when the chances of conception are low, female chimpanzees and baboons neither arouse much sexual interest in males nor exhibit much interest of their own in sex.

The sexual swellings of bonobos are different from those of chimpanzees because, to put it simply, they cover more days per cycle.[5] The result is not only that female bonobos remain sexually active and sexually attractive to males longer, but also that their ovulations, the most opportune times for conception, are concealed. In contrast to male chimpanzees, bonobo males have fewer cues when they unconsciously weigh the risks of initiating an aggressive attack in an attempt to monopolize a mate. More often than not, bonobo males content themselves with sharing copulations instead of interfering with those of their kin.

I do not mean to imply that copulations among chimpanzees always occur in aggressive contexts, or even that they are always confined to conceptive periods. Indeed, male chimpanzees have been known to peacefully share access to receptive females, and to copulate frequently during other than fertile times. But more prolonged sexual receptivity and attractiveness to males gives bonobo females more freedom than other female primates whose estrous periods are more distinctly defined. Not that other female primates are passive accomplices in male sexual conquests: we know that female chimpanzees and baboons, among others, are masters at maneuvering around males trying to monopolize their attention, and that they are clever about creating opportunities for trysts with preferred partners. We also have examples from genera in every taxonomic group of females forming lasting friendships with particular males, who appear to trade the protection they afford by their presence for the future prospects of a chance to mate. Bonobo females also use sexual contact to reassure one another. As a result, even though they disperse, female bonobos end up establishing friendly alliances with one another, and they count on their female allies for support against males.

Presumably other female primates with more restricted breeding seasons, such as muriquis, would have fewer opportunities to manipulate males with sex. Yet when endocrinologist Toni Ziegler and I examined the rainy-season copulations of muriquis, we found that they too engage in frequent extraovulatory sex.

Plotting observed sexual interactions and subsequent birth records onto measures of monthly rainfall and food availability is a standard procedure for defining the general parameters of primate reproductive seasonality. Still, without corresponding hormonal data to map onto observed copulations, there is no way to distinguish between reproductive and nonreproductive sex for the vast majority of primates who lack sexual swellings to flag their fertility.

It took some time to work out ways to obtain the necessary hormonal data from wild muriquis. Getting daily blood samples, or even urine samples, from such social, arboreal creatures would have been highly impractical. Instead we used feces, which drop in discrete deposits and are comparatively easy to collect. Like blood and urine, feces contain detectable levels of the ovarian hormones progesterone and estradiol that define female reproductive cycles, and thus provide a convenient, noninvasive way to calibrate when matings take place relative to female ovarian cycles.

With these new techniques, we were able to confirm our suspicion that the onset of muriqui sexual activity coincides with the resumption of ovarian cycling during the annual rainy season, and that females tend to stop copulating once they conceive. Even more exciting was our discovery that nearly 50 percent of muriqui matings occur during the intervals between ovulations, when the probability of conception is low (see Figure 3.5). In

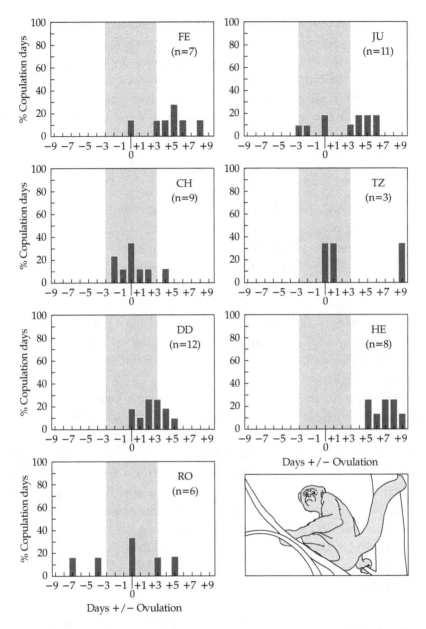

Figure 3.5 Distribution of copulation days during the periovulatory (shaded) and nonperiovulatory periods of wild female muriqui monkeys. Periovulatory periods were determined from analyses of seven female fecal steroid profiles. Copulation days (*n* given for each female) outside of the conservatively estimated periovulatory period are interpreted as being unlikely to result in conceptions.

other words, even though female sexual attractiveness is limited to the annual breeding season, that attractiveness transcends the ovulatory cycles to a greater degree than in other seasonal breeders. By mating with multiple males between their seasonal ovulations, muriqui females may be muddling the trade-offs of overt competition for males in a way that is similar to what bonobo females accomplish with their extended sexual receptivity year-round.

Confusing Conceptions

In most primates, including humans, ovulation is only detectable from the olfactory and behavioral signals sent by females. According to some accounts, modern men have the ability to discriminate between ovulating and nonovulating women, and they state preferences for an ovulator's odors above others in blind, smell-only studies. Most other male primates have better-developed senses of smell than humans do and therefore possess even greater acuity when it comes to identifying a potential partner's reproductive condition. It is not that other male primates are immune to females who solicit them for sex outside of their fertile periods, but rather that human males may be easier to confuse.

There are many different reasons why females may benefit from confusing males about their ovulatory condition, some of which apply as aptly to human reproductive arrangements as they do to other primates. Recall the tempering effects of kinship on aggression. Males are less likely to harm their own relatives, especially their offspring. At the same time, a male will benefit in an evolutionary sense if an offspring sired by a competitor is eliminated and replaced with one of his own. The possibility of paternity should therefore act as an inhibitor of aggression against infants.

Confusing paternity to mitigate male aggression is thought to be a common female tactic. In the early 1980s Sarah Hrdy pulled together the known accounts of female primate sexual and reproductive strategies. She found a surprising wealth of evidence indicating that few female primates are passive about permitting undesirable males to mate with them, or about limiting themselves to single mates when multiple partners are available. In Hanuman langurs, one of the most familiar species of colobines, pregnant and lactating females sometimes solicit sex from males who have just taken over their troops. Under such circumstances, males have a tendency to attack and kill young infants. Neither expectant mothers nor new mothers still nursing their infants are ovulating when they invite these strangers to mate, which led Hrdy to hypothesize that by confusing the question of conception, these females are protecting their offspring from male attacks.

Comparable cases of nonreproductive sex have cropped up in other species of primates in which groups of females are subjected to similar types of takeovers by males. Whenever males are physically larger than females, and therefore pose potential threats to mothers and their infants, it makes evolutionary sense that any tactics adopted by females to confuse males about their paternity would lower the risk to their infants.[6]

Functions of Fidelity

The benefits of introducing such paternity uncertainties make less sense when females have little to fear from males. For example, philopatric males already have a genetic stake in offspring sired by their kin, and should therefore be less inclined to harm infants than males whose reproductive competitors are unrelated to themselves.

Kinship among males is one form of insurance that protects infants from male aggressive attacks, and fidelity is the premium some female primates pay to gain the greatest protection from males. Male contributions to infant care take a variety of forms, from defending territories or food resources that benefit females and their offspring; to providing protection against aggression from other unrelated individuals; to carrying, grooming, or sharing food with an infant. Of course, males presumably are disinclined to make these kinds of investments, which take away from the time and energy they could instead devote to finding additional mates, unless high female fidelity increases the probability of paternity.

The New World callitrichids, which include marmosets and tamarins, have carried the cooperative care of offspring to an extreme. Like many other male primates, when male callitrichids disperse in search of reproductive opportunities, they tend to do so in pairs. In these monkeys, however, it is not uncommon to find families in which sexually mature sons and brothers forfeit their own reproductive activities to help rear the infants whom their fathers or dominant brothers have sired.

It is important to note that kinship in these and other primates (gorillas for one) is not a necessary condition for cooperation among males. Dominant males may tolerate unrelated subordinates, who bide their time as nonreproductive group members when options for finding females elsewhere are limited.

The proximity of other males, whether or not they are kin, creates convenient opportunities for female infidelities. Not surprisingly, then, one way of ensuring a female's fidelity is to keep her away from other males. In humans, concealed ovulation and continuous sexual receptivity create problems for enforcing female fidelity; as a result, we have developed uniquely cultural solutions. Mildred Dickemann has interpreted the human practice of claustration and purdah, in which women chosen to be wives or concubines of upper-caste men are isolated from the company of other fertile males, as an example of how concern with female fidelity and the measures that men take to protect it coincide with the degree to which men control resources that females and their families need. In other primates, too, obtaining protection or access to resources that some males are better able than others to provide can lead females to forgo promiscuity and instead offer sexual fidelity (or at least the appearance of it) to gain male investment in themselves and their offspring.

Extra-group Copulations

Sexual fidelity and its accompanying expectations of male parental investment have been linked to the formation of long-term pair bonds in humans and other animals. Many birds and some mammals, including primates, live in social groups composed of a single adult male and female and their offspring. For years researchers assumed that females in these groups mate exclusively with their male counterparts, thereby maintaining male investment in territorial defense and parental care. It turns out, however, that these monogamous arrangements are not all that they seem. First came reports from experimental studies in which pair-bonded male birds were captured, castrated, and then returned to their territories and mates. The altered males were biologically incapable of achieving fertilizations, yet most of the females nonetheless managed to reproduce with the help of males other than their mates. More recent genetic paternity tests conducted on a wide variety of bird species have confirmed that extra-pair fertilizations are far from rare.

Few comparable genetic data are available from wild primates, but observations of females mating with males other than the ones with whom they live are widespread. White-handed gibbons in Thailand live in family groups and defend a joint territory, just like so-called pair-bonded birds.

And, just like socially monogamous birds, both male and female gibbons pursue extra-pair copulations.

The primatologist Marina Cords was one of the first to raise questions about incongruities in the social and reproductive functions of primate social groups after observing extra-group copulations in forest-dwelling guenons, a diverse cercopithecine genus in which matrilineal females often live in social groups with a single adult male. Since then, extra-group matings have been reported in nearly all primates studied, ranging from the opportunities created by the influxes of males that appear during the brief breeding seasons of lemurs and patas monkeys, to those resulting from encounters between neighboring groups of Costa Rican capuchins.

Demography may turn out to be a determining factor in female infidelities. For example, high population densities tend to lead to more frequent intergroup encounters, and therefore more frequent infidelities. Similarly, the impetus to seek mates outside the group should be greater when the sex ratios of groups become highly skewed, or when the proportion of close relatives of the opposite sex in a group becomes high. In muriquis, both increased population density and the maturation of philopatric sons can account for the increases we have witnessed in both intergroup encounter rates and the frequency of extra-group copulations involving mothers who avoid mating with their sexually active sons.

We do not yet know whether conceptions result from muriqui extra-group copulations, because no noninvasive paternity tests have been developed for this species. However, analyzing the DNA found in hair follicles collected from chimpanzee nests in the Taï National Forest in the Ivory Coast, Pascal Gagneaux and his colleagues had some startling results. Only 6 of 13 cases of paternity could be traced to male residents in the chimpanzee community that the females frequented. The remaining fertilizations could only have occurred when ovulating females eluded both their chimpanzee consorts and their human observers to slip away for assignations with males residing in other communities.

Preventing female primates (or birds, for that matter) from pursuing extra-group partners is a problem common to pair-bonded or philopatric males alike. Human infidelities are typically attributed to love or passion, and in many cultures the costs of being caught can be high. Indeed, homicides committed by cuckolded men are generally judged more permissively than other kinds of crimes. But just as surely as the roots of human male sexual jealousy come down to the same questions of paternity that plague other male animals, so too the roots of female infidelities can be traced to practical considerations. The most compelling range from ensuring fertilizations to hedging bets on the qualities of males that their offspring will inherit, or on the resources, including protection, that potential fathers are likely to provide.

Of course, these possible benefits must be weighed against the costs to females—like the risks of being fertilized by genetically inferior males or of being injured in attacks initiated by dominant males whose aggression they incite. In fact, fidelity is a phenomenon that seems to occur only when males can impose it on females, either directly through their greater social power or brute force, or indirectly through controlling the resources that females and their offspring require.

In sexually egalitarian societies like those of bonobos and muriquis, females have both less to fear from males and more equal access to resources than do most other female primates. The fact that the most liberal attitudes toward human sexuality are found in societies where women have similarly independent social status and economic means is entirely consistent with the broader patterns across primates. And as in the case of male kin bonding, such patterns have little to do with phylogeny.

Primate Behavioral Diversity

Chimpanzees and bonobos, like all of the world's surviving primates, live under very different conditions nowadays than they did in the evolutionary past. Many primates occupy habitats that have been altered by humans to varying degrees, and their populations have been adversely affected by hunting, habitat fragmentation, or both. Behavioral traits that hold constant across widespread, isolated populations are the strongest candidates for claims about evolutionary antiquity. Yet identifying these traits requires access to more than the small handful of well-studied populations on which most of our present knowledge is based. During the 1990s pieces of the puzzle of chimpanzee societies began to fall into place, for long-term studies of wild populations across Africa are well established. Together with the insights gained from observing social colonies of chimpanzees in captivity, we have a much firmer grasp of what chimpanzee behavior is all about than we do for many primates, including other apes.

Scientific access to comparative populations of bonobos, which in the wild are restricted to a small area of the Congo, is compounded by problematic human politics in the region as well as by threats of logging. For equally endangered apes and monkeys, the opportunities to accumulate insights into the extent of both interspecific and intraspecific variability are vanishing too quickly.

Deciphering primate behavioral diversity requires the use of comparative approaches to identify the range of variation found within species and the patterns that distinguish one species from another. When apes are viewed alongside other primates in this way, we can begin to see what is—and what is not—special about them.

4

The Ape's Gift:
Meat-eating, Meat-sharing,
and Human Evolution

Craig B. Stanford

Meat is highly prized by both humans and chimpanzees. Data on cooperative hunting by male chimpanzees allow us to revisit the old Man-the-
Hunter debate, initiated when only humans were thought to be carnivorous. As a part of political and mating strategies, hunting may have driven
the evolution of social intelligence.

The form usually taken is that of the gift generously offered;
but the accompanying behaviour is formal pretence and
social deception, while the transaction itself is based on
obligation and economic self-interest.

—*Marcel Mauss*, The Gift

THE FRENCH SOCIOLOGIST Marcel Mauss, writing in the early twentieth century about traditional human societies, recognized that sharing one's most prized resources with another is not done solely because of
an altruistic state of mind. That animals transfer resources among themselves for a variety of reasons has been a focus of behavioral research for
decades. Mothers, of course, share their most vital resources—energy and
time—with their dependent offspring to ensure that they grow successfully to maturity, thereby increasing the mother's reproductive success.
But among social animals, other group members share as well.

To willingly give up hard-won food is paradoxical if an animal's goal is
its own individual survival and reproduction, often at the expense of other
group members. The relationships that emerge from sharing, especially
those involving the barter of food for something else, such as an alliance,
require a keen mind. The sharer must have a sharp memory to be able to
remember alliance networks. At the same time, the resources involved can
become the focus of a web of social interactions (see Figure 4.1). In human
evolution, complex ways of sharing key resources may have accompanied
the recent escalation of brain size and accompanying cognitive advances.
Many archaeologists believe that meat became an increasingly important
part of the early human diet at the same time that the size of the cerebral
cortex began its rapid evolutionary expansion.

The most prized of all foods is the live prey captured by predators.
Social carnivores such as lions and wolves hunt cooperatively for prey.

Figure 4.1 The politics of sharing. Alpha-male chimpanzee Wilkie shares a red colobus carcass with his male and female allies and with swollen females. Such sharing has been shown to be most liberal as males rise to alpha rank, and to decrease rapidly thereafter. (Photo by Craig B. Stanford.)

Omnivores, such as humans and some nonhuman primates, also hunt for meat, but make a different decision: whether to hunt at all on a given day. Since the diet of traditional tropical and subtropical foraging people (hunter-gatherers) and that of even the most carnivorous nonhuman primates, contains only a small fraction of meat, most omnivores spend the majority of their time foraging for plant foods. The attempt to acquire meat consists of costs and benefits, and meat's nutritional value—protein, fat, and calories—must be weighed against the costs of obtaining it—search time, energy, and risks of failure and of injury during capture. Even though the percentage of meat in the diet of both humans and nonhuman primates may be small, primates still occupy major ecological roles as predators in tropical and subtropical forest ecosystems.

In this chapter I examine the relationship between two hunting primates in particular—chimpanzees and humans—and their prey. I do so in order to infer patterns of meat-eating and especially meat-sharing that may have occurred in our earliest human ancestors. We use chimpanzees as models because among the four great apes (the gorilla, bonobo, and orangutan being the others) it is chimpanzees who hunt avidly and consume quantities of meat in some study sites that rival the meat intake of modern traditional human foragers. In considering the hunting and meat-sharing behaviors of these two primates and their implications for understanding the human fossil record, I also attempt to reframe the infamous Man the Hunter debate over the role that meat is likely to have played in the evolution of human cognition.

The Hunting Apes

Most species of nonhuman primates eat animal protein, mainly in the form of insects and other invertebrates. Only a few of the higher primates eat other mammals on a regular basis. In the New World, capuchins of the genus *Cebus* are voracious predators on a variety of smaller animals, including squirrels and immature coatis. Baboons can also be eager hunters of small mammals such as hares and antelope fawns, and in at least one site meat-eating by baboons was as frequent as for any nonhuman primate population recorded.

Only in one great ape do we see the sort of systematic predatory behavior that we believe began to be a part of our early hominid ancestors' behavior between 2 million and 3 million years ago. Chimpanzees occur across equatorial Africa, and in all forests in which they have been studied intensively they prey on a variety of vertebrates, including other mammals. At least 35 species of vertebrate prey have been recorded in the diet of wild chimpanzees, some of which can weigh as much as 20 kilograms. One prey species, the red colobus, is the most frequently eaten prey in all forests in which chimpanzees and colobus occur together. In some years chimpanzees in Gombe National Park, Tanzania, kill more than 800 kilograms of prey biomass, most of it in the form of red colobus.

In 1960, when Jane Goodall began her now-famous study of the chimpanzees of Gombe, it was thought that chimpanzees were strict herbivores. When Goodall first reported meat-eating behavior, many people were skeptical and claimed that meat was not a natural part of the chimpanzee diet. Today hunting by the Gombe chimpanzees has been well documented. It has also been observed at most other sites in Africa where chimpanzees have been studied, including Mahale Mountains National

Park (also in Tanzania) and Taï National Park in Ivory Coast. At Gombe chimpanzees may kill and eat more than 150 small- and medium-sized animals such as monkeys, wild pigs, and small antelopes each year.

Chimpanzee society is called fission-fusion polygyny and features little cohesive group structure apart from mothers and their infants. Instead, temporary subgroupings called parties come together and separate throughout the day. These parties vary in size, in relation to the abundance and distribution of the food supply and the presence of estrous females (who serve as magnets for males), so the size of hunting parties varies from a single chimpanzee to as many as 35. Party size and composition influence both the timing of hunts and the odds of success.

After four decades of research on the hunting behavior of chimpanzees at Gombe, we know a great deal about their predatory patterns.[1] Red colobus account for more than 80 percent of the prey items eaten. But Gombe chimpanzees do not select the colobus they will kill randomly, for infant and juvenile colobus are caught in greater proportion than their availability; 75 percent of all colobus killed are immature. Chimpanzees are largely fruit eaters, and meat constitutes only about 3 percent of the time they spent eating overall, less than in nearly all human societies. Adult and adolescent males do most of the hunting, making about 90 percent of the kills recorded at Gombe over the past decade. Females also hunt, though more often they receive a share of meat from the male who either captured the meat or stole it from the captor. Although both male and female chimpanzees sometimes hunt by themselves, most hunts are social. In other species of hunting animals, cooperation among hunters is positively correlated with greater success rates, thus promoting the evolution of cooperative behavior. In both Gombe and the Taï Forest, a strong positive correlation exists between the number of hunters and the odds of a successful hunt.

Hunting by chimpanzees tends to be seasonal. At Gombe nearly 40 percent of colobus kills occur in the dry-season months of August and September. This pattern is actually less strongly seasonal than in the Mahale Mountains, where 60 percent of kills occur in a two-month period in the early wet season. Why do chimpanzees hunt more often in some months than in others? It is a crucial question, because studies of early hominid diets by John Speth and others have suggested that meat-eating occurred most often in the dry season, the same time that meat-eating peaks among Gombe chimpanzees. And the amount of meat eaten, even though it constitutes a small percentage of the chimpanzee diet, is substantial.

I estimate that in some years the 45 chimpanzees of the Kasakela community at Gombe kill and consume hundreds of kilograms of prey biomass of all species. This amount is far greater than most previous estimates of the weight of live animals eaten by chimpanzees. During the peak dry-season months, the estimated per-capita meat intake is about 65 grams of meat per day for each adult chimpanzee. This meat intake approaches that of members of some human foraging societies in the lean months of the year. Chimpanzee dietary strategies may thus approximate those of human hunter-gatherers more closely than previously imagined.

Several other aspects of chimpanzee predatory behavior are noteworthy. First, although most successful hunts result in the kill of a single colobus, in some hunts two to seven colobus may be killed. Multiple kills happen frequently—21 times in 1990 alone. I estimate that from 1990 through 1993 the colobus kills of the male chimpanzee Frodo alone eliminated about 10 percent of the colobus in the home range of the Gombe chimpanzees. The likelihood of multiple kills is tied directly to the number of hunters in the hunting party. At Gombe the percentage of multiple kills rose markedly during the late 1980s and early 1990s, which meant that many more colobus overall were being eaten in the late 1980s than five years earlier. The most likely cause is changes in the age and sex composition of the chimpanzee community. The number of adult and adolescent male chimpanzees rose from five to twelve over the 1980s, owing to the large number of young males who were maturing and taking their places in hunting parties.

In the early years of her research, Jane Goodall noted that the Gombe chimpanzees tend to go on "hunting crazes," during which they hunt almost daily and kill large numbers of monkeys and other prey. The most intense hunting binge we have seen occurred in the dry season of 1990. From late June through early September, a period of 68 days, the chimpanzees were observed to kill 71 colobus in 47 hunts. Because this is the *observed* total, the actual total may be one-third greater. During this time the chimpanzees may have killed more than 10 percent of the entire colobus population within their hunting range.

To try to solve the binge question, my colleague Janette Wallis and I examined the database of hunts recorded over the past decade to determine which social or environmental factors coincided with hunting binges. Knowing that hunting was seasonal helped; we expected binges to occur mainly in the dry season, and this proved to be the case. But other interesting correlations emerged as well. Periods of intense hunting tended to be times when the size of chimpanzee foraging parties was very large; this corresponded to the direct relationship between party size and both hunting frequency and success rate. Additionally, hunting binges occurred especially when female chimpanzees with sexual swellings were traveling with the hunting party. When one or more swollen females were present, the odds of a hunt occurring were substantially greater, although party size and number of males present were also significant predictors. This co-occurrence of party size, presence of swollen females, and hunting frequency led me to ask the basic question, "Why do chimpanzees hunt?"

Reasons for Hunting

Hunting by wild chimpanzees appears to have both a nutritional and a social basis. Understanding when and why chimpanzees choose to undertake a hunt of colobus rather than simply continue to forage for fruits and leaves, even though the hunt involves risk of injury from colobus canine teeth and a substantial possibility of failure to catch anything, has been a major goal of my research (see Figure 4.2).

In his pioneering study of Gombe chimpanzee predatory behavior in the 1960s, Geza Teleki considered hunting to have a strong social basis. Some early researchers had said that hunting by chimpanzees might be a form of social display, in which a male chimpanzee tries to demonstrate his prowess to other members of the community. In the 1970s, Richard Wrangham conducted the first systematic study of chimpanzee behavioral ecology at Gombe and concluded that predation by chimpanzees was nutritionally based, but that some aspects of the behavior were not well explained by nutritional needs alone. More recently, Toshisada Nishida and his colleagues in the Mahale Mountains chimpanzee research project reported that the alpha there, Ntologi, used captured meat as a political tool to be withheld from rivals and doled out to allies. And William McGrew has shown that those female Gombe chimpanzees who receive generous shares of meat after a kill have more surviving offspring, suggesting a reproductive benefit tied to meat-eating.

My own preconception was that hunting must be nutritionally based. After all, meat from monkeys and other prey would be a package of protein, fat, and calories hard for any plant food to equal. I therefore examined the relationship between the capture success and the amount of meat available with different numbers of hunters in relation to each hunter's expected payoff in meat obtained. That is, when are the time, energy, and risk (the costs) involved in hunting worth the potential benefits, and when therefore should a chimpanzee decide to join or not to join a hunting party? And how do these compare to the costs and benefits of foraging for plant foods?

These analyses are still under way because of the difficulty of learning the nutritional components of the many plant foods in the chimpanzees' diverse diet, but the preliminary results are surprising. I expected that as the number of hunters increased, the amount of meat available for each hunter would also increase. Such a correlation would have explained the social nature of hunting by Gombe chimpanzees. If the amount of meat available per hunter declined with increasing hunting party size (because each hunter got smaller portions as party size increased), then it would be a better investment of time and energy to hunt alone than to join a party.

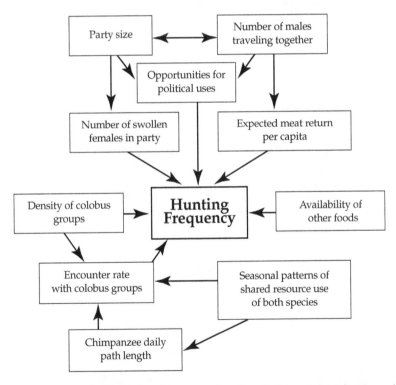

Figure 4.2 Factors that influence chimpanzee hunting decisions. Social factors such as party size are in the top half of the chart; these have been found to be positively correlated with hunting frequency based on multiple regression analyses. Ecological factors are in the bottom half of the chart. (From Stanford, 1998a. Reprinted with permission of Harvard University Press.)

As it turned out, the hunting success rates of lone hunters was only about 30 percent, while that of parties with ten or more hunters was nearly 100 percent. There was no relationship, either positive or negative, between the number of hunters and the amount of meat available per capita (see Figure 4.3). Perhaps the reason was that even though the likelihood of success increases with more hunters in the party, the most frequently caught prey animal is a 1-kilogram baby colobus. Whether shared among four hunters or fourteen, such a small package does not provide anyone with much meat.

One chimpanzee hunting party can decimate a group of red colobus in a matter of minutes. What is the likely long-term effect of intensive chimpanzee predation on the colobus population? From information on the size, age, and sex composition of red colobus groups, combined with knowledge of the hunting patterns of Gombe chimpanzees, it is possible to estimate the impact of predation on the colobus. Based on my monitoring of five colobus groups over the past four years, plus an ongoing census of a number of other groups that occupy the 18 square kilometers of the chim-

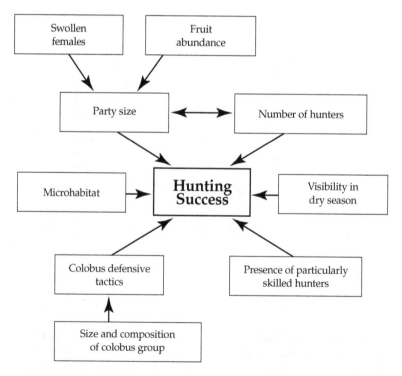

Figure 4.3 Social and ecological factors that influence chimpanzee hunting success. (From Stanford, 1998a. Reprinted with permission of Harvard University Press.)

panzees' hunting range, about 500 colobus live in the chimpanzees' range. I estimate that approximately 75 to 175 colobus are killed by chimpanzees each year. The annual mortality rate in the colobus population that is due to chimpanzee predation is thus between 15 and 35 percent, depending on the frequency of hunting that year.[2]

The Earliest Hominids

During the 1980s some archaeologists spoke of a "scavenging niche" occupied by Pliocene hominids. Early-genus *Homo* and australopithecines were thought to have a made a living by locating the carcasses of the large savanna ungulates (Figure 4.4). One model even suggested that *Homo habilis* relied extensively on the leftovers from the kills of saber-toothed cats for its subsistence. The argument goes that among the early hominids of the Pliocene era, the cognitive capacities for cooperatively hunting and for sharing of meat were not likely to have been present.

We know from both the chimpanzee data and from accumulating paleoanthropological data that this view is almost certainly wrong. Although it is true that stone tools—clear evidence of butchering of large mammal prey—only appear in the archaeological record from 2.5 million years ago, before that time hominids of many species were probably eating meat opportunistically. This meat would have been obtained through both hunting and scavenging whatever sources of animal protein could be found—by capturing prey or by pirating meat away from a carnivore's kill.

This view of early hominids is relatively new. They were not bold biggame hunters, as earlier thinkers such as Raymond Dart and Sherwood Washburn imagined. Nor were they bipedal weaklings, shuffling about the savanna in search of carcasses from which they tried to steal scraps of meat without being caught by a lion or leopard. Instead, these earliest hominids—who we now believe lived mainly in forested rather than open grassland habitats—were skilled, efficient hunters of small animals and also engaged in what Henry Bunn calls power-scavenging, taking meat from carnivores by driving them away from their kills. This view of the behavior of our earliest human ancestors fits much better with current data than do previous views.[3]

Was the mode of getting meat mainly scavenging or hunting? The dichotomy has existed since the earliest writings on the subject. For the most part, it is a false dichotomy. In the present day, practically no animals hunt but refuse to scavenge, nor do mammals scavenge but never hunt (vul-

Figure 4.4 Early hominids probably foraged for meat in whatever form was available. Hunting for small mammals, combined with the active pirating of carcasses seen here, would have yielded substantial animal protein and fat in the diet, even at the earliest stages of hominid evolution. (Drawing by Sarah Landry, 1975; prepared in consultation with F. Clark Howell for *Sociobiology* by Edward O. Wilson. Reprinted with permission of Harvard University Press.)

tures are the best-known pure scavengers among birds). We traditionally contrast lions and hyenas, for example, as the hunter versus the scavenger, when in fact both species engage extensively in both foraging modes. Lions, in fact, frequently power-scavenge meat away from hyena kills. Thus any scenario for the meat-eating habits of a hominid ancestor should begin with a dual hunting/scavenging mode of subsistence.

Chimpanzees obtain very little meat by scavenging. In nearly 40 years of observation at Gombe, the total number of passive scavenging events, in which carcasses are found and eaten from while chimpanzees are foraging through the forest, is less than 20. There are only a few records of scaveng-

ing from the Mahale Mountains. Wild chimpanzees often treat fresh carcasses as though they are inedible; on one occasion Gombe chimpanzees
climbed inside the body cavity of a freshly killed bushbuck (probably the
victim of a leopard attack) and rolled around as if in a playhouse. The carcass aroused intense curiosity but not a hunger response, in spite of more
than 75 readily available kilograms of protein and fat.

Why chimpanzees do not scavenge is unclear; there are relatively few
opportunities for finding carcasses in a forest compared to a more open
habitat. African forests lack the large herds of ungulates and other larger
mammals; the biomass of these animals is lower than that of animals in a
savanna. We know that cultural traditions are crucial to understanding
many aspects of chimpanzee behavior; perhaps without a steady supply of
carcasses the tradition of treating dead animals as prey never became established. The lack of scavenging by modern apes, combined with the
near-ubiquity of hunting of live prey by even the most ardent scavenger
species, make it highly unlikely that early hominids at any stage of human
evolution restricted themselves to scavenging. The portrait of hominids as
obligate scavengers is unsupportable, at least in light of the chimpanzee
evidence. The question relative to the early human diet has clearly become
not whether early humans hunted or scavenged, but how much of each
behavior contributed to the diet of each species during each geologic time
period.

Problems with *Man the Hunter*

Why It Was Wrong . . .

The traditional dichotomy between hunting and scavenging goes directly
to the heart of our perception of what it means to be human. To be a hunter
implies a number of qualities: physical strength, agility, the ability to cooperate and coordinate actions—and probably to communicate. To be a scavenger, an early hominid would need few of these traits. In a now-infamous
1968 volume called *Man the Hunter*, Sherwood Washburn and Chet Lancaster laid out a scenario for the evolution of human cognition. In an oft-
quoted statement, Washburn and Lancaster considered hunting to be at
the heart of human uniqueness: "In a very real sense our intellect, interests,
emotions, and basic social life—all are evolutionary products of the success of the hunting adaptation."

Man the Hunter was, of course, all wrong. It attempted to explain why men evolved large brains and social communication skills: because the hunting adaptation placed a high premium on coordinative and communicative abilities. In many taxa of animals, from hawks and eagles to the social carnivores, social hunting has evolved. Neocortical brain size and elaboration are not necessary attributes in any of them. The idea that social hunters need complex systems of communication comes from watching human hunters, whose success rate (at least when using traditional weapons) does not necessarily exceed that of small-brained social predators like cheetahs, lions, and crowned eagles. The Harris hawk of the deserts in the American Southwest is a highly cooperative hunter, preying in pairs or in family groups on rabbits and rodents. It does so with a cerebral cortex not very different, as far as we know, from the brain of other birds. The ability to coordinate predatory behavior with the parallel actions of groupmates is therefore not linked phylogenetically to the expansion of the brain's neocortex.

Man the Hunter never addressed the question of why women also evolved. Because of its failure to take the female sex into account in human evolutionary models, the book created a backlash against sexism that was centuries overdue in the study of humanity. Whole schools of anthropological thought coalesced, partly in reaction to the blatant failure of Washburn and Lancaster even to consider women in the human evolutionary process. Female biological anthropologists in turn have taken *Man the Hunter* apart, analyzed its approach, and asked what role females played in human origins.

Adrienne Zihlman was among the first critics, pointing out that even though Washburn and others had focused on male hunting as the presumptive central behavior of our ancestors, models based on modern hunter-gatherer people offer little support for hunting as the most essential mode of subsistence. In fact, among many hunter-gatherer groups women collect most of the animal protein through their less glamorous but more consistently productive gathering of small animals and other foods. Moreover, the diet of most tropical foraging peoples is only in small portion meat. These two observations suggest that male-centered views of hominid origins erred by omitting a large part of the behavioral ecology of the Hominidae.

The backlash against *Man the Hunter* permeated not only feminist cultural anthropology; it may also have influenced behavioral biology. Coming at a time when more women scientists were entering the field of behavioral science, *Man the Hunter* pushed scientists to reinterpret the foundations of the societies of primates and other animals. Accumulating field data had begun to show that females of many social species were anything but the passive receptacles of male mating ambitions that they had long been assumed to be. Instead, females came to be seen as active strategists, differing from males in their means of attaining reproductive success but equally intent on doing so. We now view most primate societies as female centered rather than male controlled, based on the strength of the female bonds present. Females are also the "ecological" sex. Their own reproductive gains are more intimately tied to the physical environment than those of males, because of offspring-rearing needs. This fact, combined with the lower reproductive potential of females, leads females to be discriminating and choosy about whom they mate with, and places them firmly at the center of the mating system of many species.

. . . and Why It Was Right

In spite of a generation of anthropologists who tried to ignore it, *Man the Hunter* was fundamentally correct in its focus on meat-eating (but not hunting) as the heart of the Plio-Pleistocene human adaptation. Opponents such as Zihlman focused correctly on the nutritional importance of the female role in gathering protein, but ignored the undeniable fact that the role of meat in human society has never been merely nutritional. Even among chimpanzees the reasons underlying the capture of prey are social as well as nutritional. Detractors of *Man the Hunter* pointed out that meat is not a calorically valuable resource in most modern tropical foraging societies. Despite its nutritional benefits, the unreliability of its capture combined with women's efficiency at gathering other sources of animal protein render it somewhat superfluous. However, in all human societies from forager to pastoralist to farmer, meat is a highly valued food resource, accorded a status far beyond its nutritional worth. It also plays a central role in sexual division of labor and is often a commodity used by males for the patriarchal control and manipulation of females. A collected pile of roots and tubers may provide a family with nutrients, but anthropologists have long noted that these plant foods do not evoke nearly the excitement, nor generate the exchange of goods and services, that meat does.

Meat provides a social as well as political motivation for the hunter, among human hunter-gatherers as well as chimpanzees. Chimpanzees at Gombe spend hours in pursuit of colobus prey, even though palm nuts are readily available and would provide more calories per gram, more calories per minute of foraging, and more saturated fat per gram than monkey meat does. The return rate for most members of a Gombe chimpanzee hunting party is very low—a few scraps of the carcass or even just drops of blood—even though an hour or more is often invested in the hunt, and foraging for other plant foods is forgone while waiting for a potential prize. Males use meat to secure and maintain political alliances, to publicly snub rivals, and at times to attract estrous females. At Gombe, low-ranking male chimpanzees who capture colobus become magnets for swollen females, with whom they share meat and copulate until discovered by a higher-ranking male. The dominant animal may then appropriate the meat and engage in the same meat and sexual politics as his predecessor. I also saw male chimpanzees capture colobus and then discard them to rejoin the hunt, as though the point was the capture rather than the meal.

Once a kill is made, the carcass is likely to become the focus of intense political activity. We see cultural diversity from one wild chimpanzee population to the next in the pattern of sharing that follows. Gombe chimpanzees are utterly nepotistic and Machiavellian in their use of the carcass; captors share mainly with their family members, allies and swollen females. In Taï, hunters receive a share of meat regardless of the captor if they have participated in the hunt.

The failure of other apes to eat and use meat provides an instructive counterpoint that supports a nonnutritive basis for hunting. The chimpanzee's closest living relative is the bonobo. Wild bonobos hunt much less frequently than chimpanzees do, and not because of a lack of monkey prey or an inability to catch them: researchers have witnessed bonobos capturing monkeys, carrying them about as playthings for hours, and then discarding them without any interest in the "prey" as something edible. Moreover, when meat is obtained it is often controlled by females. Male bonobos may be less inclined to hunt than male chimpanzees not because meat benefits them any less nutritionally, but because their kill is usually lost to females. These apes, often cited as counterpoints to the chimpanzee pattern of male domination and political control of females, may demonstrate that meat is sought by chimpanzees for social reasons *primarily,* since females control carcasses and thereby eliminate the males' incentive for acquiring them.

The variety of reasons for hunting animals in human societies is far greater than in nonhuman animals. Among foraging people, however, many of the cost-benefit equations for time and energy invested in hunting should be similar to what we see in nonhuman hunters. Unlike social carnivores, both humans and apes must decide each day whether to forage for meat at all, and if they do, where and how to capture it. For hunter-gatherers, meat can be a pursuit more or less costly depending on the season, the species sought, the weapons employed, and the habitat in which the prey are apt to be encountered.

One of the most notable differences between the structure of human forager and chimpanzee societies is the importance of status and hierarchy. Human foragers are famous in ethnographic accounts for their lack of obvious leadership or overt political authority; they are typically described as egalitarian. Chimpanzees meanwhile are extremely hierarchical, with all adult males dominating all females plus adolescents. A male chimpanzee's social career is a determined climb up a dominance hierarchy, through an intricate web of political alliances and conspiracies. This dominance comes into play at times of meat-sharing in some chimpanzee populations.

In human forager societies, if enhancing the nutrition of oneself and one's family is not the only impetus for hunting, what else is? A number of anthropologists have asked this question, with intriguing results. Kim Hill and Kristen Hawkes examined the hunting patterns of Achè hunter-gatherers in the Amazonian forests of Paraguay. They found that the Achè attempt to balance the energy and time expended on obtaining food during foraging trips (the Achè are no longer true foragers; they are settled on missions, from which they embark on foraging trips) with the calories they obtain from the foods they hunt and gather. Their foraging decisions take into account the caloric and nutrient value of a variety of forest foods and the expected return rate on the effort and time needed to procure them. Meat is a major part of the Achè diet (as much as 45 percent of the food mass acquired), which leads Hill and Hawkes to ask why the Achè ever bother to gather plant foods. The most invoked explanation for this is risk reduction; the high risk of failure inherent in hunting is mitigated by collecting other foods.

Hawkes suggests that we consider the way Achè men share the meat they capture in order to understand why it is hunted so avidly in the first place. Among chimpanzees, kills are sometimes solicitously guarded and at other times are shared in highly strategic ways. Among many human foragers, kills are brought back to camp and distributed liberally to families other than the hunter's own. In fact, the captor himself most often does not control the meat or its distribution after capture; "ownership" is not necessarily linked to hunting success. Having one's catch consumed by families other than one's own may mean a type of risk reduction is going on, a hedge against the day when one's own family might go hungry without a reciprocating gift of meat from another hunter.[4]

A paradox may be seen in this sharing pattern, however. Even though some hunters contribute much larger quantities of meat than other hunters do to the hunting party for general distribution, there is no relationship between the amount of meat a hunter contributes and the size of the share of meat he may receive for his own family later in a time of need. This fact is perplexing; we should expect some kind of reciprocity if meat-sharing is about reducing the risk of future hunger. Hawkes argues that the pattern is elegant evidence for a nonnutritional role of hunting; that is, the function of hunting by the Achè may not be entirely nutritional. Through their hunting men are seeking the availability of valued goods that are widely shared and therefore offer the opportunity for politically clever uses of the meat, to secure alliance and even to obtain more sexual liaisons. In other words, Hawkes reasons that Achè men use meat to show off their prowess as hunters. Direct control of the resource is unimportant, and the fact that carcass distribution is not controlled is not a disincentive for men to hunt. Showing off by obtaining a public resource that all may use may enhance a man's status. It may also enable him to increase his reproductive success; Hillard Kaplan and Kim Hill showed that those men who were the best hunters were also named by women as the preferred partners in extramarital relationships. Thus, the sharing of meat by Achè men may be considered part of a mating and political striving strategy rather than simply the provisioning of one's family.

The strengths of the "showing-off" model for hunting and meat-sharing are threefold. First, it provides an explanation for why men hunt at all while women gather. Vital animal protein gathered by women, even in ample quantities, would not diminish the desire of men to capture prey, owing to the social impetus for sharing. Second, the model may explain why some men contribute more to sharing networks than others, based on individual variation in political and reproductive goals. Third, it explains why large game is pursued: the largesse it provides is the sort of meat bonanza that a successful hunter can effectively use for social strategizing. Hawkes's sexually selected model parallels my own model of chimpanzee meat-eating as sexually selected, if hunters in both species obtain more mates as a result of being successful hunters.

To Share Is Human

Why would a captor of meat have any desire whatsoever to share the carcass rather than consume the bounty himself? Many theories have been advanced. The earliest studies of meat-sharing by humans and other animals assumed that individuals shared for the good of other group members, and to promote general harmony. We know today that such altruistic behavior is difficult to maintain in the face of opportunities for selfish behavior. The common behavioral denominator even for social animals is individual nutrition, survival and reproduction. Generous sharing that carries no self-directed benefit is a paradox, as first noted by the evolutionary biologist Robert Trivers. It is resolved by Darwinian theory only if we hypothesize that the favor will be returned at a later date, either by being shared with the giver or through some other social currency such as support in a conflict or perhaps by being groomed.

Sharing is common among social animals, as recently reviewed by Bruce Winterhalder. Vampire bats, for instance, engage in extensive reciprocal sharing that is crucial to their survival. Gerald Wilkinson showed that these bats often fail to find their normal meal of mammalian blood, and mortality would be extremely high if not for blood they receive as shared meals regurgitated by other, often unrelated, bats in the colony. Blood sharers form alliances in which the partners regurgitate food back and forth. It is a classically reciprocal altruistic relationship, predicated on long-term relationships, mutual need, and ease of resource transfer. Similarly, Bernd Heinrich has studied meat-sharing among ravens; these birds share information about the location of the carcasses of the deer and moose that are their primary wintertime food sources. Ravens call unrelated ravens to inform them of the presence of a carcass. They appear to share information readily, rather than hoard their knowledge of the meat and the meat itself, both because the cost of sharing is extremely low and because the benefits of a reciprocal system in which one reaps the reward of information later in exchange for giving it now, are great.

These two examples of sharing behavior reflect the idea that cooperation occurs mainly when major selfish interests are at stake. Craig Packer and his colleagues showed that when a lion pride's food (savanna ungulates) is abundant, meat consumption is unrelated to group size. But when food is scarce, those lions that either forage alone or else hunt cooperatively in groups of five or six achieve the greatest per-capita consumption of meat.

Moderate-sized hunting parties of two to four reduce intake. The same relationship between the number of hunters and their meat consumption was shown by Kim Hill and Kristen Hawkes to occur as in Achè foragers, and I found a similar result in my own work on chimpanzee predatory behavior. It seems that individuals may adjust their pattern of association in order to maximize their odds of obtaining meat for themselves.

Darwinian explanations for sharing therefore focus on the selfish incentive for what often appears to be generous altruism. Nicholas Blurton-Jones first proposed that when a hunter-gatherer controls a carcass, he may allow others to take meat from it simply because the energy expended in preventing the theft would be greater than the loss of the bit of meat itself. This model, known as tolerated theft, can occur when the carcass is big enough to be divisible. For it is then subject to diminishing marginal value as its pieces are taken and the consumer becomes sated. A second model, risk-sensitive sharing, is the hedge against shortfalls that we examined earlier. Altruistically reciprocal (tit-for-tat) sharing occurs when resources are exchanged for one another.

These explanations for sharing, along with entirely social theories such as Hawkes's showing-off hypothesis, are relevant to both human forager groups and chimpanzee society. The political importance of intelligence in chimpanzee society has been well demonstrated, and its ties to exchange and barter are evident. Frans de Waal's *Chimpanzee Politics* crystallized our view of the manipulative and strategic nature of chimpanzee social intelligence, although decades earlier Robert Yerkes had described exchanges of sex and food among captive chimpanzees. At Taï, Christophe Boesch has documented highly cooperative hunting and meat-sharing after a kill that rewards the chimpanzees who participated in the hunt. The Mahale Mountains M-group alpha-male chimpanzee used meat to affirm his newly won status at the top of the hierarchy, a pattern that declined with his tenure as alpha. Such a barter system can commonly exchange goods for services: de Waal, for instance, showed that captive chimpanzees shared food in exchange for grooming.

When males control meat or any other highly valued resource, opportunities abound for use of the resource as a manipulative tool. Robin Dunbar and others have proposed that the increase in the size of the brain's neocortex came as a result of the advantages of social intelligence, the ability to navigate the icebergs of one's political network to obtain selfish benefits such as status and mating success. Even those acts that appear to be altruistic and performed to increase harmony among individuals occur mainly when they enhance one's own social situation as well. Richard Byrne observed that those primate species that engage in what we consider to be high-level forms of social strategizing such as deceit ("tactical deception") are also those taxa that possess the largest brains. Social, Machiavellian intelligence is currently the explanation for human brain evolution with the most support.[5]

Origins of Patriarchy, Origins of Equality

Those who control the distribution of highly valued resources control the lives of others. When men control valued resources, the course of human history has shown that they control the lives and even the reproduction of women. It is unimportant that the controllable resource in question—meat—may not be objectively worth its value in terms of calories and fat. It is relevant only that the resource is accorded high status by both possessor and potential recipients. When this situation exists in human societies, power flows backward to the controller of the resource; a "gift" becomes a tool of barter that ultimately holds hostage those in need of the gift.

We are all familiar with one manifestation of this sort of social system. It is called patriarchy. Obtaining resources, sharing and bartering resources, and controlling distribution of resources—in these three areas, males in both human and chimpanzee societies tend to strive for and occupy controlling positions. This feature is not often discussed by human evolutionary scholars because it may seem to denigrate the role of women in human prehistory. Yet as we strive to achieve a view of early human societies that is equally rich in detail for females and males, we also strive to achieve a view that is accurate and that owes its details to factual evidence and not to our Western sense of balanced-gender ethics. For generations the role of women in the human evolutionary process was largely ignored; a Paleolithic glass ceiling existed. The situation has changed, much to the betterment of evolutionary anthropology. Still, we must not be so excessively swayed by Western values of gender equality that our portrait of early humanity is as distorted in its own way as the views of earlier generations—a goal more easily stated than attained.

Both the advocates and the critics of *Man the Hunter* missed their mark by focusing on the capture of meat rather than its control. The capture of meat requires highly evolved hunting adaptations, both morphological and behavioral, but not cognitive. Advanced cognitive abilities come into play only after the kill has been made and a carcass lies waiting for consumption and distribution. Moreover, in both hunting and scavenging the end product is a carcass that may be divided. The dichotomy of hunted prey versus scavenged carcass becomes peripheral to the key question of what sharing may have meant to the evolution of social cognition.

It is nearly a cultural universal that males engage in patriarchal behavior. The occurrence of matriarchal "Amazon" societies of ruling women in the human past is now considered a myth, and even in the most egalitarian societies men exert political power over women that is manifest in a variety of ways. Women, of course, also have spheres of influence, but rarely do they involve the direct control or distribution of the most valued foods eaten by the group. Obtaining meat is therefore a goal of both men and women, often for different reasons. Men may seek game in order to use meat for tactical socializing, while women value meat for its amino acids while they are gestating or lactating. Since males control meat, women may need to be tactically clever in order to acquire a share. This reasoning, more than the ideas expressed in *Man the Hunter,* is likely to have been an evolutionary impetus for the expansion of human cognitive skills that accompanied increasing quantities of meat in the early human diet.

It may be time, therefore, to resurrect Glynn Isaac's sharing model of the 1970s, in which key events of human evolution happened because meat (whether obtained by scavenging or by hunting) provides a social currency which, united with the social brain hypothesis, accords well with current paleoanthropological, primatological, and human-forager data.

Finally, we should consider what happens to this relationship between meat, sharing, male domination, and cognition when males do *not* control resources. This is a rare circumstance, but it does occur. Among bonobos, females form coalitions that keep males from dominating them in a variety of social situations. Bonobos appear to be less aggressive hunters than chimpanzees are, although differing field conditions and a history of habituation make it difficult to know with certainty. When bonobos obtain meat (usually duiker antelope), it is most often controlled by females, not males. Females have even been reported to take meat from males with impunity. This sort of behavior is unimaginable in chimpanzees. Under conditions of female coalitional behavior, sexual equality in political power may emerge. The result of course is a very different set of social relationships among male-dominated primates, and it also creates a whole new arena for using cognition. Differences between chimpanzees and bonobos in social behavior may also involve different types of tactical socializing employed by each sex under very divergent pressures.

Reframing *Man the Hunter* in terms of meat-sharing and social intelligence has profound implications for how we view our deep ancestry. Our interpretations of the fossil record, of hunter-gatherer societies, and of great-ape societies may to some extent reflect Western values about gender roles. These roles are under greater scrutiny today than at any time in human history. Males and females in both human and nonhuman primate societies have a strong vested interest in obtaining key foods, but both the means and the goal of obtaining the food differ between the sexes. Studying these differences leads us to many of the most interesting features of our humanity and of the origins of human cognition.

5

Out of the *Pan,* Into the Fire: How Our Ancestors' Evolution Depended on What They Ate

Richard W. Wrangham

What allowed the descendants of chimpanzee-like ancestors, 5 million to 6 million years ago, to colonize savanna woodlands? They appear to have depended on eating roots. About 4 million years later, our ancestors became human. A new hypothesis suggests that this happened as a result of the adoption of cooking.

PREHISTORY is a murky time, full of imposed fantasies and imagined worlds, darker and less penetrable the farther back we go. But the distant past is illuminated by two periods when our ancestors evolved especially fast—the only major transitions that link humanity to the African apes. These give us a way to think about the mystery of hominization, the subject of this chapter. How did ape become human, and what did the changes mean for the evolution of social behavior?

The first big change, 5 million to 6 million years ago, was from a forest ape to an australopith. On the far side of that evolutionary bridge, we can reconstruct our ancestor as a quadrupedal, knucklewalking ape that looked much like a living chimpanzee. Still unnamed and undiscovered, this last representative of the preaustralopiths is best thought of as a member of the genus *Pan*. For convenience, it needs a name. So let me be cheeky and supply one. *"Pan prior"* seems reasonable: the early chimpanzee.

When one population of *Pan prior* evolved into an australopith, in many ways it did not change. Among the characteristics retained from its forest ape ancestor were long arms that rotated at the shoulder, a small female body (about the same mass as a chimpanzee), a capacious gut, large jaws, and a characteristically ape-sized brain. So members of this new species presumably still climbed and armhung in fruit trees; compared to humans, they still ate relatively low-quality food (that is, of low caloric density); and they still thought essentially as living apes do. No fossils have yet been discovered or named.

Although the australopiths remained similar in several ways to living apes, they were placed in a new taxonomic tribe because they experienced three substantial changes in their evolution from *Pan prior*. First, unlike *Pan prior*, they lived outside the rain forests. Second, their teeth were no longer adapted for shearing and chewing foliage; instead, they were adapted for crushing. Finally, the hallmark change was toward bipedal locomotion.

Based on their hind limb bones, it is probable that from the beginning the earliest australopiths walked upright often, perhaps habitually. The change in life style signified by these three changes is considered to be so important that the australopiths are credited with being the first hominins—a group composed of humans and our bipedal relatives.[1]

The australopiths have sometimes been called ape-men. As we shall see, it is certainly reasonable to think of them as having the qualities of apes because the sizes of their brains, bodies, guts, and teeth were essentially apelike. As for being semihuman, their only significantly humanlike trait was bipedalism. Does this warrant their being called ape-men? It is a matter of opinion. Most people think of the important traits of humanity as being our large brains, as well as our small jaws, guts, and teeth; the large bodies of our females; and our commitment to terrestrial locomotion. Since these characteristics are not found in the australopiths, I think our hominin pioneers are better called *apes* than *ape-men*. To distinguish the australopiths from the living (forest) apes, I call them woodland apes, although bipedal apes would be an equally appropriate name.

For the next 5 million years or so, until 1.4–1.2 million years ago, the woodland apes thrived. They produced at least three genera (*Ardipithecus, Australopithecus,* and *Paranthropus*) and a variety of species—perhaps eight or more. Whether two or more species ever shared the same habitat is unknown, but it is certainly possible. The earliest species, *Ardipithecus ramidus,* had rather thin enamel on its teeth, but subsequently all the members of this varied group retained the three principal signals of their common ancestor (woodland habitats, large and thick-enameled molars, and bipedal locomotion). They flourished in the African woodlands much as gorillas, chimpanzees, and bonobos have flourished in the African rain forests. As is true of the forest apes, the ecological adaptations, social systems, and cultural traditions of the woodland apes were probably diverse and fascinating. Unfortunately, we know almost nothing about the varied life styles of these species. What we do know is that along the way one population did something extraordinary. In a second major transition, it became human.

Hominization is evidenced by two sets of fossils dated between 2.5 million and 1.9 million years ago. The first is a highly variable fossil species that fits the concept of a "missing link" (or "ape-person") better than any of our ancestors. This is the famous "handy man," traditionally called *Homo habilis.* But *Homo* means "human": how human were they? They had somewhat larger brains than their predecessors—about 550 cubic centimeters, compared to 410–515 cubic centimeters for other australopiths. They were humanlike also in showing evidence of eating large mammals, to judge from the ancient cutting tools that sometimes accompanied their remains. These flakes of stone enabled the hominins to slice meat off the bone, leaving cut marks still visible on the fossils of species such as extinct antelopes.

Despite brain expansion and eating of large mammals, *habilis* remained strikingly like other australopiths in most aspects. They had ape-sized females, they were skillful climbers, they had big mouths and large teeth, their males were much bigger than their females, and they grew in the ape manner—more quickly than humans. These fossils were a mixture, therefore, which suggested to the paleoanthropologist Philip Tobias (who first reluctantly named them *Homo*) that the ideal moniker would be *Australopithecus–Homo habilis*. Although Tobias' idea would have captured nicely the intermediate nature of this species, the rules of taxonomy do not allow a name that combines two genera. Because in most respects *habilis* now looks like other australopiths, the anthropologists Bernard Wood and Mark Collard have argued that we should call them prehuman: *Australopithecus habilis* rather than *Homo habilis*.

The next step was the big one. Around or before 1.88 million years ago, *Australopithecus habilis* evolved into *Homo erectus* in a transition that might have lasted but a few tens of thousands of years—or less. Compared to *Australopithecus habilis*, its *Homo* descendants had bigger females (estimated to be a full 60 percent heavier), smaller teeth, smaller guts, arms no longer adapted for hanging in trees, and longer legs. They also experienced a larger rise in brain size than previously seen, almost doubling their brain volume to over 1,000 cubic centimeters—well on the way to the 1,355 cubic centimeter value for living humans. Furthermore, their males were no longer hugely bigger than females: the degree of sexual dimorphism in body size fell dramatically to become indistinguishable from the pattern in modern *Homo sapiens*, where males average a mere 15 percent heavier than females. These ancestors of ours were essentially the same size and shape as modern humans, albeit smaller brained and with more robust bodies. For the first time, they could be put in clothes and given hats, and they could walk down a New York street without generating too many stares.

The "gradualist" anthropologist Milford Wolpoff is so struck by the magnitude of this change from *Australopithecus habilis* that he calls this new species *Homo sapiens*. But most scientists prefer *Homo erectus* (the name I use here), reserving *sapiens* for the populations that emerged between

200,000 and 150,000 years ago. Whatever the species name, all agree on the genus: by this time our ancestors were sufficiently modern in body size, body shape, locomotion, and diet that they deserved to be called *Homo,* or human. Though important changes were still to come, such as continuing increases in brain size, reductions in robustness, and increased cultural complexity, they were smaller, steadier transitions.

In the rest of this chapter I try to link apes with humans by reconstructing the natural history and social lives of the transitional species. Of course, scenarios that aim for coherence and plausibility are as vulnerable as they are ambitious. Though the fossil evidence grows continuously and is now quite extensive, the kinds of questions we ask and the demands we make of it grow correspondingly. The puzzle is still taking shape.

The Origin of the Australopiths

I began by suggesting that the undiscovered 6-million-year ancestor of the australopiths was so chimpanzeelike that we can call it a member of the chimpanzee genus—hence, *Pan prior.* This claim comes from the combination of two remarkable pieces of evidence.

Considered alone, the first point is unsurprising. Gorillas, chimpanzees, and bonobos are more similar to each other in many aspects of their morphology than to any other species. Examples are their thin-enameled teeth, evidently adapted for slicing and chewing vegetative foods such as leaves and herbaceous stems, and their knucklewalking locomotion, such that these apes place their weight on the knuckles of their hands rather than on their palms.

Because these traits are restricted to the African apes (not being found in any of their more distant relatives), similarities like these indicate that the three species had a common ancestor with the same essential features. For example, their nearest relatives are orangutans, who have thick-enameled teeth and who put their weight on their fists instead of on their knuckles when they walk on the ground. Admittedly, it is theoretically possible that instead of being caused by shared descent, these similarities evolved independently in the different species. Knucklewalking might have emerged independently in gorillas, on the one hand, and in chimpanzees and bonobos, on the other, whereas their common ancestor could have been a species of ape that did not walk on its knuckles. But there are many such similarities, for example in basic growth patterns. Therefore, as Colin

Groves has shown, we can be rather confident that the similarities among gorillas, chimpanzees, and bonobos reflect a common origin.

If we accept this argument, the genetic evidence becomes highly informative. Surprisingly, it slides humans right into the middle of the African ape group (see Figure 5.1). Gene comparisons tell us that the ancestors of gorillas branched off 6 million to 8 million years ago. Later, at around 5 million to 6 million years, the line giving rise to the australopiths (and later to humans) split from the line giving rise to the modern chimpanzees and bonobos. In other words, chimpanzees and bonobos are more closely related to humans than they are to gorillas. Yet the morphological similarities, as we have seen, are best attributed to the descent of all three from an ancestor that had the same similarities. This means, as the paleontologist David Pilbeam has stressed, that the last common ancestor of chimpanzees and humans must have had the traits shared by gorillas and chimpanzees. Therefore, our *Pan prior* was most likely a black-haired, big-jawed, arm-hanging, knucklewalking species that ate seasonally available soft ripe rain-forest fruits. To judge from the commonalities among the living apes,

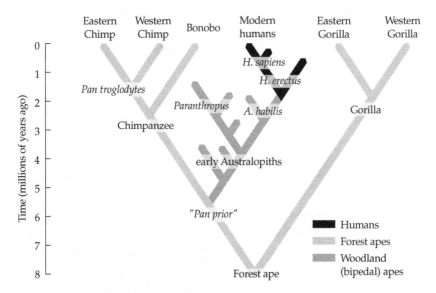

Figure 5.1 A simplified evolutionary tree of the family Hominidae. The hominids are shown clustered in three main types of adaptation (forest apes, woodland apes, and humans). The pattern of branching among the hominins (woodland apes and humans) was certainly more complicated than is indicated. Further fossil discoveries will clarify the picture.

it ate additional leaves and piths when fruits were hard to find, and could not have survived without access to rain forests, or at least to forests in permanently watered gullies, which had the ripe fruits and soft foliage necessary for daily subsistence.[2]

The Lives of Australopiths

Habitat

The australopiths were diverse. They spanned at least 3.0 million years (from *Ardipithecus* at 4.4 million years ago to *Paranthropus* at 1.4 million–1.2 million years ago), and more likely, they lived for as much as 4 million or 5 million years (assuming an earlier split from the forest apes). They occupied almost the length of Africa, from Chad and Ethiopia to South Africa. They included some species that were relatively gracile (mostly in the earlier phase of their evolution) and others that were relatively robust. And the fossil assemblages with which they were associated included animals from tree-dwellers to grass-eaters, so these woodland apes must have lived in a variety of habitats, as Richard Potts has emphasized. But there was a limit. Specifically, there is no evidence that they occupied closed-canopy rain forests, those that were sufficiently well watered to support apes like the African species alive today.

Several points suggest this limitation. First, none of the australopith fossils have been found in association with forest apes, even though chimpanzee and gorilla ancestors must have been present in Africa throughout the australopith era. Second, australopith fossils are always associated with at least some mammals indicative of seasonally dry areas, such as mole-rats, giraffes, and grass-eating ungulates. Third, the sites of australopith fossils are in locations that today are some of the driest in Africa, from the Sahara and the Rift Valley to southern Africa.

Trees must have been important to these apes, however, to judge both from the fossil association of australopiths with arboreal animals, such as fruit bats, and with pollen from trees; and from the evidence of their upper bodies, which were shaped for climbing or for hanging by one arm in the manner of modern apes. Some australopith habitats have been reconstructed as "subtropical forests" and would have contained many large trees bearing edible fruits. Others would have been more bushy, containing seasonally available foods such as young leaves, immature seeds, and sweet flowers. Since all probably offered opportunities for climbing, their habitats are best called woodlands.

What, then, enabled the australopiths to make the transition from rain forest to these more seasonal habitats? To answer this question, consider why modern apes do not occupy woodlands (habitats with less than about 1,200 millimeters of rain per year).

The most obvious constraint is food. Dry-country chimpanzees invariably eat fruits from permanently watered trees for a substantial portion of the year. When fruits are scarce, they can eat seeds from pod-bearing trees found in woodlands, but these too are only seasonally productive. When fruits and seeds are both scarce, leaves and stems become important for survival. But in woodland habitats that have no rain forest, such foliage appears, like soft fruits, to be too scarce year-round to allow survival of modern apes.

Diet

Something novel happened, therefore, to enable the australopith descendants of chimpanzeelike ancestors to colonize the woodlands fringing the rain forests, and then to speciate repeatedly. Although bipedalism was doubtless important, an increased ability to process woodland foods must have been essential. Most people assume that the dietary trick that allowed australopiths to colonize the woodlands can be diagnosed from their extraordinary teeth, which are exceptionally large and thick enameled compared to those in humans or any living ape.

Only a limited range of major food types could have sustained these woodland apes. We can exclude ripe fruits, seeds, and flowers with some confidence, because their availability would have been seasonal (as they are even in rain forest), preventing them from being reliable as fallback foods. Leaves and stems are also unlikely candidates, not only because fewer soft foods of these types would be available compared to the rain forest, but also because the thickening of australopith tooth enamel compared to the forest apes meant that they no longer produced the shearing edges needed for cutting foliage. Meat has often been considered a potentially key food, but it is an unlikely candidate for these apes, partly because (again) the morphology of the teeth looks wrong. Meat-eaters tend to have cutting teeth, whereas those of australopiths were blunt.

One class of foods, however, satisfies our search. Underground storage organs (USOs), such as tubers, corms, rhizomes, and other roots, appear ideally suited to explain the woodland ape teeth and habitats. Although USOs always occur at low densities in rain forests, they are predictably abundant in more open areas (where herbs benefit from storing water and nutrients during dry periods). Few animals eat them, because USOs are hard to dig up and are often tough or rich in toxins. For many modern human foragers, however, they are a staple of the diet, becoming more important when high-quality foods such as fruits, seeds, honey, and meat are scarce. We know that USOs have a long history in Africa, because for at least 20 million years mole-rats that specialize on eating USOs have been present. Indeed, root-eating mole-rats were found in the fossil assemblages that include *Ardipithecus ramidus*. Larger mammals that eat USOs are few, but they include various species of pigs, which were abundant in australopith habitats; and, strikingly, the morphology of pig molars shows strong convergence with hominin molars. Digging tools of horn cores and bone have been found in association with *Paranthropus* in South Africa. Finally, there are many different types of USOs, sufficient to allow ecological divergence in the apes exploiting them. For example, some woodland apes may have specialized on small USOs pulled from the marshy edges of lakes, while others may have focused on digging out large roots from among rocks on hillsides (see Figure 5.2).

In all such cases, a crucial aspect of USOs is that they would have been available as fallback foods, offering a means to survival when preferred foods were scarce. Mobile shoulders and other signs of climbing suggest that these woodland apes still foraged in trees when conditions were favorable. But like the three living African forest apes, it was probably on the ground that the woodland apes found their fallback foods during periods of fruit shortage. By this view, the ability of gorillas, chimpanzees, and bonobos to find and chew foliage as a fallback food was replaced in the woodland apes by an ability to obtain and eat USOs. This was a critical adaptation enabling the woodland apes to survive periods when natural selection was at its most intense.

One scenario for the initial speciation is the so-called East Side Story. By this account, originating from Yves Coppens and Adriaan Kortlandt, a population of ancestral apes was isolated east of the Great Rift Valley during an especially dry period, unable to cross the supposedly treeless valley and therefore kept out of contact with the main, central African ape population. Some geographic barrier or isolating mechanism is indeed required to explain how any population evolves independently of the rest of its species.

But events as extreme as the East Side Story are unnecessary. If *Pan prior* were restricted to forests, less specific isolating mechanisms can be easily envisaged without invoking a particular geographic feature such as the Great Rift Valley. Thus, periodic climatic changes undoubtedly led to alternate expansions and contractions of the range of the central African rain forest. Mute witness is given to such changes nowadays by the occurrence of certain forest species in isolated patches kilometers from the central bloc. Witness the populations of red colobus in distant mountain ranges, riparian systems, and islands of eastern Africa.

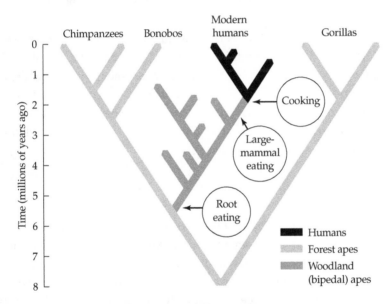

Figure 5.2 The crucial influence of three dietary innovations on the social lives of our hominin ancestors. Exactly when they were adopted is a matter of debate. The key advance that opened woodland habitats to occupation by apes may have been the ability to exploit roots and other underground storage organs, such as bulbs. According to this idea, it was root-eating forest apes who became the australopiths, the group of woodland hominins who then radiated into a variety of nonrain-forest habitats throughout eastern and southern Africa. By 2.5 million years ago, at least one species of this group was using flakes of stone as knives to cut meat from the bones of large mammals (such as antelope, hippopotamus, or elephants). More than half a million years later, rapid evolutionary change produced the first humans. The adoption of cooking (of both vegetables and meat) is a plausible explanation.

Ancestral populations of *Pan prior* were probably often isolated in such forests, whether north, east, or south of the equatorial forest belt. When these forests died out, most of the time the apes would have died out too, having neither sufficient fruits nor sufficient leaves to sustain them in the woodland during the dry season. But *Pan prior* probably had behavioral traditions that varied among populations, much as we see in chimpanzees. Perhaps a population acquired some root-eating traditions prior to losing its forest, like the chimpanzee population known from the eastern Congo that digs for large tubers among the cinder blocks of its volcanic soil. Perhaps it already used digging-sticks, like the Senegalese chimpanzees who use them to search for buried water. Or maybe it found itself in an environment with easily acquired USOs, much like the marshy herbs eaten by gorillas in the Central African Republic and the Democratic Republic of Congo.

The point of such speculation is not to suggest that a particular mechanism can be confidently identified, but merely to show that the steps do not appear prohibitive. The regular expansions and contractions of the central African forest bloc, over a cycle of tens of thousands of years, would have offered repeated opportunities for fringe populations of *Pan prior* to adapt to drier habitats, rather than go extinct. One such population, it appears, responded to the challenge.

Locomotion

Equipped with stronger upper bodies and better grasping feet than humans, the woodland apes were obviously expert tree-climbers. Their mobile shoulders show that they could have fed in outer branches the way living apes do, sitting while supported by one hand holding a branch above. Like chimpanzees, in trees they presumably ate seasonally available fruits, seeds, honey, and perhaps other more specialized items such as occasional flowers or social insects.

It seems reasonable also to think that they would have liked to sleep in trees. All the great apes make night nests, important for protection from predators. Would the kind of trees preferred by sleeping apes have been sufficiently common in the australopith woodlands? In the dry season, when many trees would have shed their leaves, there may well have been problems. Even in the relatively well forested habitat of Gombe, appropriate sleeping trees are sufficiently scarce in the dry season that chimpanzees often reuse old nests. Just as savanna baboons are nowadays restricted to living in areas where either cliffs or trees afford them nightly refuges, so the woodland apes may have been restricted to areas with sufficient safe sleeping sites.

Sleeping and frequently feeding in trees, the australopiths nevertheless were terrestrial travelers, like the forest apes. The difference between forest apes like chimpanzees and woodland apes like *Australopithecus* is that the woodland apes habitually walked upright. Oddly, no one knows why. This hallmark trait of the hominins remains a mystery, puzzling because the first attempts must surely have been awkward and inefficient.[3]

The difficulty of imagining how the initial clumsiness of bipedal walking was overcome is compounded by the fact that even when skilled at walking, bipedalists would have been more vulnerable to predators than when on all fours. Humans are very slow sprinters compared to African apes and other large mammals. The same was doubtless true of australopiths, because bipedalism eliminates the galloping action of a quadruped. If nothing else changed, therefore, bipedalism would have made the hominins more vulnerable to predation than before.

In a parallel way, we have an intriguing hint that bipedalism was related to predation risk in another ape. The Miocene ape *Oreopithecus* from Tuscany and Sardinia was once believed to be quadrupedal. Analysis of postcranial materials by Köhler and Moyà-Solà, however, suggests that it was bipedal. Interestingly, the faunal assemblage with which it is associated shows signs of insular dwarfing, that is, size reduction of a type associated with isolation on a predator-free island. The implication: escape from predators made bipedalism possible for *Oreopithecus*, because in the absence of predators being slow incurred little cost.

The australopiths, by contrast to *Oreopithecus*, clearly lived in a dangerous environment, which at different times and places included not only close relatives of modern predators such as lions, but also carnivorous bears and saber-toothed cats. The lesson suggested by *Oreopithecus* is that the australopiths were able to become bipedal only because they found a way to lower the risk of predation. In addition to escaping into trees, therefore, they are apt to have had other forms of defense, such as stabbing weapons, thrown rocks, swung thorn-branches, or cooperative attack. Perhaps their digging tools gave some defense by causing charging predators to impale themselves. Speculative as such ideas are, it is hard to imagine these woodland apes evolving their slow new form of locomotion without some way of reducing the threat from predators.

These woodland apes did not escape predation completely, as shown by an australopith skull punctured with holes that match the canines of a fossil leopard. Furthermore, in keeping with expectations if they were vulnerable to predators, they appear to have had life-history parameters similar to those of the modern forest apes. That is, compared to humans, they matured early and died young. Subject to ambush in thick vegetation, constrained by bipedalism to a relatively slow escape on the ground, these woodland apes could hardly have led easy lives.

Groups

In eastern Ethiopia savanna baboons and hamadryas baboons interbreed. These two incipient species are thought to have become separated 300,000–400,000 years ago, before later meeting again. There are few differences in their bones, and none that would indicate to a primatologist of the future how these species differed in their social lives. Yet the differences are large. Hamadryas live in fission-fusion groups, within which exclusive mating units interact with one another through alliances of adult males, leaving females largely powerless. Savanna baboons live in stable groups, with no exclusive pair bonds but with intragroup relationships strongly influenced by important alliances among adult females. If all that remained of these species were their fossils, it would be difficult indeed to reconstruct these differences. In a similar way, the woodland apes will probably forever conceal much of their diversity in social behavior.

Although we cannot be too ambitious, we can imagine a few traits. For example, Robin Dunbar (see Chapter 7) shows that among living primates, brain size is broadly related to the size of the social communities, the networks of individuals who interact peaceably. The brains of woodland apes were marginally larger than those of the forest apes, a relationship which suggests that their communities were at least as large as those of chimpanzees and bonobos.

Similar evidence comes from bipedalism. Because human day-ranges are unusually long (around 10 kilometers for male foragers) compared to apes (longest in chimpanzees, at 3–5 kilometers), it seems likely that bipedalism may extend the length of day-ranges. Because the woodland apes were bipedal walkers, they probably also had long day-ranges, further expected because the relatively unproductive habitats they occupied, with preferred food sources distributed even more patchily than in the rain forest, would have put a premium on efficient travel. This hypothesis gives support to the idea of large communities—perhaps, like chimpanzees, of up to 100 or more numbers.

Within communities, did australopiths forage in stable groups or in temporary parties? We do not know. Most primate groups, such as those of mountain gorillas, are stable; the same set of individuals sleep, eat, and travel together every day. This arrangement has advantages not only because it reduces the risk of predation, but also because it allows individuals to form alliances to defend themselves against aggressive members of their own species. Stable groups are a luxury that not all species can afford, however: food scarcity or dispersed food supplies can force normally stable groups to split up. The food regime of chimpanzees in eastern Africa, for example, is so dispersed that the social community never meets as a complete unit.

The woodland apes could in theory have adopted the system of grouping of either the gorilla or the chimpanzee; there are arguments both ways. The danger from predation makes it hard to imagine a mother feeling comfortable traversing the woodlands alone except for her six-year-old offspring and the baby she is carrying. The presence of others could help to warn and protect her. High risks from predation would therefore seem to favor large, possibly stable, groups. On the other hand, stable groups require that the woodland apes must have exploited a food source that minimized the effects of within-group competition for food. If they followed the principles that apply to living apes, their small body size suggests that they would have been more like chimpanzees, having no access to such a food source. They would therefore have formed temporary parties, smaller when preferred foods were relatively scarce.[4]

The conflict between the benefits and costs of sociality suggests a third possibility that combines aspects of the gorilla and chimpanzee systems— a mixed pattern exemplified by hamadryas baboons or human foragers. In these species, a relatively stable sleeping group breaks by day into smaller foraging parties. The mixed strategy is a tempting model, because both hamadryas baboons and human foragers live in drier habitats than their relatives, just as the woodland apes did. If the woodland apes adopted the hamadryas/human pattern, we would want to know if the smallest units of the foraging parties were lone individuals (as is sometimes the case among human foragers) or exclusive mating groups (as in hamadryas baboons, where groups average one male and two females).

Accordingly, we need to think about whether, within communities, females and males were equally gregarious. Among chimpanzees, mothers are more solitary than males and childless females. The reason is apparently that motherhood makes the females slow walkers. Babies can be exhaustingly heavy to carry, and they can also delay their mothers through their play or willfulness. As a result, travel is more costly for mothers than for other adults, in terms of both energy and time. But foraging in parties requires extra travel: the bigger a party, the more fruit trees it must visit in order to find enough food. Faced with the choice between efficient or gregarious foraging, it seems that mothers must forgo some of the benefits of sociality in order to minimize their daily travel. Therefore, if woodland apes had chimpanzeelike diets and moved between discrete food patches such as fruit trees, we can expect their mothers to have traveled alone when the fruits were poor.

In sum, a reasonable guess about the woodland ape grouping systems is that they consisted for the most part of social communities, perhaps (like chimpanzees) 20 to more than 100 strong. Concentrations of suitable trees may have been favored as sleeping areas, allowing large sleeping parties to aggregate when food was abundant. When fruits and seeds were scarce, smaller foraging parties were formed and probably created occasions for more frequent predation by carnivores. Mothers were somewhat less gregarious than males—except, of course, for males guarding their mates from other males.

Social Relationships

The anatomist Owen Lovejoy once suggested that bipedalism evolved in order to allow males to bring food to their mates, but his much-publicized idea of monogamy among the woodland apes now seems far-fetched. One of several difficulties is the question of sexual dimorphism in body size. Fossil evidence is generally interpreted as showing that male australopiths were far bigger than females—about twice the weight, much as in gorillas. This major sex difference in body mass implies that selection favored fighting ability in males, presumably because some males were much more successful than others in competing for females. Since monogamy yields little benefit to males fighting each other, it therefore seems improbable among australopiths. Other obstacles to Lovejoy's scheme include skepticism that a male who left his mate to find food for her could guard her from rival males, the absence of any evidence for home bases, the matter of why females became bipedal, and evidence that australopith life histories resemble those of apes, not of humans as Lovejoy's scheme implied they should.

In addition, no living nonhuman primates exhibit monogamy-within-social-communities. Part of the explanation is that any female who mates exclusively with a low-ranking male within a social group can expect to find her offspring the target of infanticide attempts by more dominant males. For instance, in chimpanzees and gorillas infanticide occurs regularly. Likewise, infanticide by the large males of the woodland apes was almost certainly a worrisome threat. In living species infanticide seems significant as a shaper of social relationships, and it probably was for the woodland apes too.

One kind of protection against infanticide is an exclusive mating bond. It is reasonably effective in a species such as the gorilla, where a male and female are able to spend all their time together and there are few competing males. But it cannot work for the chimpanzee, because fission-fusion foraging means that pair bonds are at best fleeting consortships, which last only a few days or weeks before the pair separates. This temporary mating bond may benefit a female by saving her from being harassed during her mating period, but it fails to protect her from infanticide by other males. Because she is sometimes forced by her foraging needs to travel alone, a female chimpanzee uses a strategy opposite to that of a gorilla. Instead of restricting paternity to a single mate, she normally gives every male she meets a chance at paternity: she mates widely, typically with all males in the community.[5] See also Chapters 1 and 2.

Among the woodland apes, therefore, females may have had a variety of options in their defense against infanticide. Another intriguing possibility is that females were able to form alliances with one another to protect themselves against males, as bonobos do. However, the evidence that males were much larger than females suggests this is unlikely.

Finally, we can approach the question of sociality from a different angle. Behavior is sometimes phylogenetically conservative. Female-bonded groups, for example, are found in all the species of Asian macaques and in all the African forest guenons that have been studied, so unstudied species of these taxa are expected to be female bonded as well. In a similar way, behaviors that are found in all the living African apes and in humans are reasonable candidates for having been present in australopiths, subject to acceptance or rejection based on the logic of behavioral ecology.

Accordingly, we can suggest that, like the African apes, adolescent female australopiths would typically have emigrated to a new community, leaving their mothers behind. They would not have fought in aggressive coalitions against one another. Males, on the other hand, can be expected to have been philopatric (residing as adults in the communities where they were born), to have competed for status, and to have formed coalitionary bonds against other cliques in the same community. They would also have given one another support in intergroup encounters that, in species having fission-fusion foraging, would have included lethal violence.[6] Neither males nor unrelated females would have helped significantly in the rearing of young, except to protect from threats such as predators and infanticidal neighbors.

Becoming Human

The Dietary Change

Though we might wish for the first signs of humanity to have come in cultural achievements, the drama is in fact more mundane. There are strong signals of a massive improvement in dietary quality between *Australopithecus habilis* and *Homo erectus*. These include the 60 percent increase in female body mass, a large reduction in tooth and jaw size, and (to judge from the reduced flaring of the lower ribs) a substantial reduction in the size of the gut. Additionally, this period witnessed a continuing increase in brain volume beyond the size of the living African apes. *Homo* seems to have eaten better than *Australopithecus*. Why is that?

For several decades the main hypothesis has been a new commitment to meat-eating—to either hunting or scavenging, or both. The idea that meat was eaten by hominins is supported by the evidence of cut marks on bones of large mammals at about 2.5 million years ago. The potential importance of meat for evolutionary change is supported by the fact that it is expected not only to improve dietary quality, but also to influence social relationships, as Craig Stanford discusses in Chapter 4.

When we think about meat-eating as an influence on hominization, a problem arises with the timing. Eating of large mammals began by 2.5 million years ago. But *Homo erectus* is not found until 1.88 million years ago, at which point it emerges rather suddenly. So meat-eating began more than half a million years earlier than hominization, a long delay that is not easily explained.

Further, in terms of the diet as a whole, no modern tropical foragers rely entirely on hunting. Plants are invariably a vital part of the food supply, as they are for chimpanzees—especially during periods of food scarcity. This is true even though the weaponry of modern humans is much more sophisticated than the stone and wood tools of early humans. Almost everyone agrees that plants have always been critical components of our ancestors' diet.

Plant foods are hard to process and digest, so much so that the teeth and digestive systems of different primate species are closely adapted to their specific plant diets. Accordingly, the changes that occurred in the digestive system when our ancestors became human should have had a considerable impact on the ability to digest plants. Other things being equal, the reduction in the size of teeth, jaws, and guts should have *reduced* our ancestors' ability to digest plant foods! It is hard to understand how selection could have favored a reduced digestive ability unless something dramatic happened to the plant component of the human diet.

Partly for this reason, my colleagues James Jones, Greg Laden, David Pilbeam, and Nancy Lou Conklin-Brittain and I have recently proposed an alternative to the meat-eating hypothesis. We suggest that cooking was even more important than additional meat in the diet. According to our hypothesis, hominization happened because a late gracile population of australopith apes—such as *A. habilis*—learned to use fire to improve the digestibility and range of its plant foods. It is possible also that cooking of meat was highly significant.

An intriguing way to consider the impact of cooking is to ask when it evolved. No present-day humans are known to live without cooking, the staples of many cultures cannot be eaten raw, and cooked food is invariably easier to digest than raw food (in other words, it is of high quality). Whenever cooking originated, therefore, it should have left strong signals of an improved diet. There turn out to be only two points in human evolution (prior to the adoption of agriculture) that signal such an improvement.

First, Loring Brace has argued that cooking was adopted about 250,000 years ago. There is evidence of earth ovens being used at that time by peoples living in the glacial zones of Europe where, Brace suggests, they needed to defrost meat. In addition, an improvement in nutrition consistent with cooking is signaled by a reduction in molar area. Against Brace's idea, the reduction in tooth size at that time was relatively modest and was the only signal of improved energy availability. It was therefore a weaker signal of dietary quality than those found at 1.9 million years ago.

Second, much evidence attests to the control of fire 300,000–400,000 years ago or earlier. For this reason, we believe that the signal offered by reduced tooth size at 250,000 years indicates not the origin of cooking, but merely an improved cooking technology.

The only other time of strong signals of improved nutrition in the hominin fossil record is around 1.9 million years ago, suggesting that cooking is most likely to have originated then. Although archaeologists are uncertain about when fire was first tamed, two studies suggest control of fire had been achieved shortly thereafter (1.7 million and 1.6 million years ago). Our calculations suggest that cooking has a stronger effect on calorie intake than an increased supply of uncooked meat, and can therefore account more easily than meat-eating for the increase in female body size at 1.9 million years ago. Because cooking caused the diet to be softer and more readily digested, it can readily account for the reduction in tooth area and gut size, as well as the increased energy needed for fueling a larger brain.

Control of fire can explain several other features of early *Homo*, too. Climbing adaptations were lost at this time, perhaps because the population that cooked also used fire to scare away predators. Hence individuals could sleep and eat more comfortably on the ground. In addition, greater security from predators would reduce mortality rates, which would have favored longer lives, slower growth, and delayed maturation.

The cooking hypothesis suffers from the lack of an archaeological consensus on when exactly cooking arose. Nonetheless, it has the merit not only of providing an alternative against which to test the meat-eating hypothesis, but also of explaining a number of previously mysterious features of hominization. The origin of cooking can be visualized as a result of australopiths picking over half-exposed roots accidentally baked by a natural fire, a scenario that fits the evidence that Africa was getting drier at about the time that *Homo* originated.

Once cooking became important, foraging economics changed because of a shift from immediate to delayed consumption. Delayed consumption implies gathering (therefore the use of gathering tools) and probably a communal return each evening to a central location (although nightly changes of camp are also possible). It further implies the likelihood of sex differences in foraging strategy, as we shall see below.

The shift from immediate to delayed consumption undoubtedly was not total. Even today human foragers routinely find edible foods during the day, which they may eat raw before returning to camp. Foraging may have been directed toward fruit or honey or meat (which could have been eaten raw), with cooked USOs as a fallback. Even if foods that could be consumed immediately continued to be important in the diet, the hypothesis is that on a daily basis the cooking of USOs became, by virtue of its high caloric impact, a core foraging adaptation.

Changes in the Mating System

Is it possible to link the cooking hypothesis to one of the most striking features of the anatomy of humans compared to African apes and other hominins, which is our reduced sexual dimorphism in body mass? Compared to many primates, men are small—a mere 15 percent heavier than women on average, raging across populations from 6 percent to 23 percent, as Richard Smith and William Jungers have shown. By comparison, male weights exceed female weights by 30–40 percent in chimpanzees and bonobos, and around 100 percent in gorillas and orangutans. Most estimates for the australopiths suggest that males were heavier than females by 50 percent or more, and probably close to 100 percent in at least *Australopithecus afarensis.*

Looking back in time, this pattern of relatively large human females, or small males (compared to the ratio of male-to-female body mass in other hominoids), is first detected among the earliest members of *Homo erectus*. One of the distinctive changes that occurred when humans evolved is that our males stopped towering over our females. This reduction in the relative size of males is of special interest because among nonhuman primates sex differences in size are related to the mating system in a systematic way that conforms to sexual selection theory.

We have already considered the logic in our discussion of australopiths. In species whose females offer few mating opportunities, variance in male mating success tends to be high. Males benefit by competing aggressively for those few chances. The largest males are favored, because in most primates bigger males tend to win fights. By contrast, in species whose females mate frequently, males tend to have more equal mating success. Rather than investing in fighting ability, therefore, such males benefit by remaining smaller, perhaps because this allows them to live longer. Among primates, these ideas have been tested by comparing species whose females vary from having less than 20 copulations per birth (gorillas, for instance) to 100 or more (chimpanzees, bonobos, and humans). As expected, species with relatively frequent mating are indeed the ones with less imposing males.

This useful result allows us to ask when the human mating system is likely to have developed into its present form. Our human system is unusual compared to that of most primates. Not only is the timing of ovulation essentially concealed from men, but women are sexually attractive throughout their menstrual cycles and mate regularly during their interbirth intervals (aside from a short period of abstinence following a birth). We are obviously a frequently mating species with low variance in male mating success.

How long has our system been in place? If human evolution has worked like the evolution of other primates, the answer is clear. Whenever our frequent-mating, low-male-variance system originated, it should have been associated with a marked decrease in sex dimorphism in body mass compared to its immediate ancestor. Only one such decrease exists in the entire course of hominin evolution. It occurred at around 1.9 million years ago with the origin of *Homo erectus*.

This strongly suggests that at this time the current human mating system essentially began. No later change is known to have occurred in sex dimorphism in body mass or in other traits relevant to fighting (such as upper-body strength). Therefore, significant changes in the mating system since then seem unlikely. Rather, at 1.9 million years ago a complete suite of mating adaptations evolved, including the concealment of ovulation, permanent female attractiveness, a high number of matings between births, and fairly equal distribution of mating opportunities among different males.

The Social System of Early Homo

Cooking and the modern human mating system should both have left a signal in the fossil record. We have seen that in both cases, the evidence points to the same time, at the origin of humanity. What link can explain the simultaneous emergence of cooking and the human mating system?

This question has not been asked until recently, because cooking was never thought to have arisen so early. Indeed, even the time when the modern form of the human mating system originated has been undecided. Others have proposed that the human sexual system arose with the earliest australopiths (around 5 million years ago), or with the evolution of hunting large mammals (around 2.5 million years ago), or with modern humans (around 0.25 million years ago).

Though the timing is uncertain, most concepts designed to explain the peculiar human mating system have associated it with male-female bonding. Two ideas have been particularly meaningful. In one, males have been viewed as effective providers, because they offered females meat. This concept has been central to hypotheses that stress the importance of male contributions from hunting and scavenging. The other idea, elaborated most recently by Sarah Mesnick (based on the work of Barbara Smuts), views males as valuable protectors because they guarded females from sexual coercion (including herding and aggression to mothers or their babies). In both scenarios, ovulation became concealed because females benefited by escaping the sexual attentions of too many males, thereby giving increased paternity confidence to the investing male.

Both of these schemes have merits, but they have problems too. The meat-provider hypothesis is undermined by data showing that in foraging societies, the meat supply of mothers is not necessarily influenced by the success or failure of her partner's success at hunting. The group has a strong cultural rule: all meat must be shared equally among the different families. Kristen Hawkes has pointed out that if early humans were similarly egalitarian, there would have been no advantage in terms of meat-getting for women to bond with particular men.

The sexual harassment theory seems to work well in contemporary society, but has not been shown to account for the origins of the system. In particular, it assumes part of what it seeks to explain—that females benefited by mating selectively with a single male. After all, if bonobo females rarely object to "sexual harassment," why should early human females have objected? We need to explain also why females did not mate widely, with all males in the group.

The link to cooking offers a third explanation, which my colleagues and I have called the theft hypothesis. It proposes that a significant effect of cooking was that food items had to be accumulated into a small area (next to a fire) and retained there for several minutes or hours. During this waiting time, the food (roots, seeds, and the like) would have been a tempting target for hungry individuals with little food in their own hands. Furthermore, dominant individuals would rapidly have learned that they did not need to gather and cook for themselves; they could live as scroungers.

At the dawn of cooking, males were perhaps 50 percent heavier than females, and presumably socially dominant over them. So it is easy to imagine that they could have stolen food from females. The situation need not have been confrontational (unless a female resisted the big male who sidled up to her hearth and helped himself to a roasting root). Other females and their offspring were potential scroungers too. Females, accordingly, would have needed to protect their hard-won food supplies as effectively as possible. Mothers would have particularly needed an effective defense system, since the welfare of their offspring would have been critically dependent on the nutrition they provided. According to the theft hypothesis, therefore, females bonded with males to protect themselves from scroungers.

If mothers benefited from having the best possible protection, they would have been in competition with one another. The theft hypothesis suggests that it was female-female competition for male food-guardians that generated the unusual human mating pattern. Females who were more sexually attractive obtained a higher quality of food-guardian. Selection thus favored mothers who "deceived" males into finding them attractive at all points of the menstrual cycle. We end up with a theft hypothesis that portrays the human family as originating in a swirl of sexual and domestic politics around a kitchen hearth!

Conquering New Worlds

In this chapter I have set aside many of the complexities that enrich the analysis of the fossil record. To those who are entranced by the puzzles, I urge a plunge into the primary literature, where a more confusing and more cautious world will become evident.

In keeping with my streamlined version, however, let us quickly trace the almost 2 million years after hominization. Soon after their rather sudden evolution from australopith apes—perhaps within 100,000 years— early humans left Africa and colonized tropical regions as far away as Java, in modern Indonesia. Because cooking would have given these early humans a substantially greater range of high-quality foods, it would have allowed them to occupy habitats inappropriate for their ape ancestors.

Over the ensuing millennia various forms of humanity came and went— including the Neanderthals, who lived in Europe and adjacent regions of Asia until some 45,000 years ago. Brain size increased and sometimes fell. Language took over. African populations colonized the rest of the world at least once again, ending in a wave of modern *Homo sapiens* around 150,000–200,000 years ago. Then, about 40,000 years ago, cultural diversity bloomed in the creation of ornaments, tools, and art. By 12,000 years ago, agriculture introduced the modern era.

Although such events were turbulent in the context of human achievement, the human frame had entered a calmer era. In comparison to the great shifts from our ape past, there has been little change for 1.9 million years in features such as body size and degree of sexual dimorphism, or shape of the foot or the shoulder, or nature of the teeth or the face. This relative conservatism of human morphology suggests an equivalent conservatism in selective pressures. In that light, for almost 2 million years all humans, from *Homo erectus* and Neanderthals to ancient and modern *Homo sapiens*, have lived under the same basic system of social ecology and sexual selection. Prometheus is said to have created humans by animating clay figures with fire. If the foraging and mating systems of humans were indeed shaped powerfully by cooking, the ancient Greek myth that attributes humanity to the gift of fire may be close to the truth. The adoption of cooking perhaps completed the framework upon which the great diversity of human behavior has been assembled.

6

Social and Technical Forms of Primate Intelligence

Richard W. Byrne

Can we reconstruct the "ancient mind" by speculating about the capacities upon which our ancestors' cognitive evolution was founded? Was it in the social domain that things began—with deception and perspective-taking; in the technical domain—with tool use; or in the context of foraging and complex food manipulation?

E VERYONE LOVES the old "intellectual development as a tree" metaphor. Youth's upward striving, every person developing in a slightly different way as he or she grows up, branching out in manifold directions to take a part in the rich interwoven canopy of human accomplishments, the flowering of the intellect that eventually comes to fruition . . . heartwarming stuff. To pursue the metaphor a bit further, how did the "trunk" of this tree arise to form the firm basis necessary to support the endless diversity of modern human cognitive achievements? That is, what are the evolutionary origins of the cognitive potential that underlies these myriad achievements?

This chapter is about these evolutionary "roots." Although they lie deep in our ancient ancestry and little tangible evidence of behavior is preserved archaeologically, I argue that we can learn a great deal about the evolution of human intelligence by using comparative evidence from living animals. (Note that I use the word *intelligence* simply as a neutral label for a species' package of information-processing capacities, equivalent to the term *cognition*.) However, the evidence has to be used carefully and in a particular way, so the chapter includes an outline of the various methods available for working out what happened in the evolution of intellectual abilities. In the past, researchers have used several different (and mutually contradictory) ways of relating animal evidence to human intellectual origins. Some were based on misguided premises and have given the whole enterprise rather a bad name, so they need to be sharply distinguished from the procedures of contemporary evolutionists.

Evolutionary Building Blocks of Intelligence

At one extreme is the assumption that our descent was linear and progressive, passing through a series of previous forms that closely resembled living species: the Great Chain of Being, in which lemurs evolved into monkeys, monkeys into apes, and apes finally (and best of all) into people. According to this delightfully simple idea, when we studied a lemur or monkey we were seeing in a direct way what our ancestors were like. Evolution was treated as progressive and additive: monkeys are lemurs-plus, apes are monkeys-plus. This progressive view links with another nineteenth-century belief, the idea that ontogeny recapitulates phylogeny. On this theory, the early development of an organism retraces the stages through which it "progressed" in evolution, so that studying development reveals evolutionary history.

Sadly, this approach is just plain wrong. Progression is a comfortable myth, but modern animals have evolved for exactly as long as we have, and often in very different directions since our lines of descent diverged. True, they are our relatives; but none of them are our ancestors. Nor can we assume that the common ancestor of monkeys and humans must have been like one or another modern species, since each line of descent will have undergone subsequent selection. Recapitulation is equally invalid, for similar reasons. A species' ontogeny is as much under the control of natural selection as any other part of its life cycle: we cannot expect a developing organism to work through its evolutionary history, handy as that would be for us. If an embryo uses a method of respiration different from the adult form, it can only be because that method is functional at that developmental phase. Development is not a quick and easy method of discovering evolutionary history.

Almost diametrically opposite to these notions, the Radical Behaviorism of the twentieth century asserts that there are no essential differences between species, since all of them simply learn; conveniently, a simple and cheap species such as the white rat could be studied instead of an awkward and expensive one such as a chimpanzee—let alone the whole range of species that would be needed for comparison. But comparison is unnecessary if all species learn the same way. A variant of this view, favored by experimental psychology, was that all animals are intellectually the same—except one. Humans alone are quite different because they possess language, which underlies every major intellectual achievement of humanity.

This discontinuity theory is implausible because evolution cannot proceed by inspired jumps, only by accretion of beneficial variants of what went before. Language is a unique yet highly complex adaptation. If it is completely unrelated to the cognition of other species, a remarkably (and improbably) rapid adaptive process must have gone on in the few million years since we last shared an ancestor with a living ape. More likely, the flowering of human intellect in language is built upon, and composed of, many other abilities—which may have histories independent of one another.

In contrast, the all-species-are-equal view of radical behaviorism *might* have been right, but in fact the evidence is against it. True, a huge range of animals have similar basic capacities to learn about events that are associated in time and space, and to remember the useful or painful consequences of their actions (called, respectively, Pavlovian and operant conditioning, in behaviorist terminology). Yet these capacities are not uniformly available: rats avoid novel food after experiencing nausea, but not after feeling an electric shock; a young monkey rapidly learns to avoid a snake, but not a daffodil, after seeing adult monkeys show fear in its presence. Moreover, some species of animal show much more powerful abilities than others to understand the world and deal with complex situations; evidence for this claim will be reviewed later in this chapter. All animals are not equal. In particular, I argue that current evidence can be summarized with four propositions:

1. Monkeys show more complex behavior than most mammals.
2. This ability results from an enlarged neocortex that allows rapid learning.
3. Great apes demonstrate some understanding of intentions and causes.
4. This comprehension is based on an ability to perceive, and to build, complex novel behavior.

The methods used to interpret such evidence in terms of human mental evolution are based on robust premises, outlined in the next section.

It may be helpful as a preliminary to appreciate that evolutionary questions can take several different sorts of answers, and to consider what a full explanation might include. First, there is an answer in terms of *function*. In the case of an evolved trait, this amounts to knowing what environmental challenge the aptitude met, how it caused increased inclusive fitness in those individuals carrying the trait (that is, how they came to leave more descendants than those that lacked it). The original environmental challenge is often referred to as the selection pressure that caused the evolution of the trait. For intelligence, this translates into asking, "What environmental challenge caused our distant ancestors to gain a fitness advantage from cognitive adaptations in particular, rather than some other advantage?"

Efficiently dealing with this challenge may or may not be the current function for which the living inheritors of the trait find it most valuable. A second question is, "Are the challenges for which our intelligence is most crucial today the same as those ancient ones that led to its evolution, or has intelligence taken over new functions?"

Next, we might consider the form of an evolved aptitude: some are very specific abilities, minutely fitted for one particular use (like the immensely long tongue of an anteater, handy for eating ants but of little use for anything else); others have multiple uses (like the grasping hands of a raccoon, effective in many situations, even sifting through the garbage from an overturned trash can). For intelligence, this contrast is equivalent to asking, "Is intelligence a set of job-specific skills (or modules), each useful in only one domain, or is it a more general and flexible capacity, employed for many specific purposes?" If intelligence is modular, we might expect our modern intellectual capacities to retain distinctive signs of the original function for which each evolved. (Note that only if intelligence is general purpose might it make sense to assess it in a single way such as an IQ test.)

Finally, we may look for an answer in terms of chronology and dates: "Are the skills we observe in modern humans recent or ancient?"

To put in perspective the contribution that evolutionary methods can make to the understanding of human psychology, consider three contributory strands to the origin of the modern human mind. One strand comes from an individual's personal history and social environment. It can only be studied by focusing on modern humans, their ontogeny and cross-cultural variation—although cultural variation in other species can inform debate about the earliest human cultures (see Chapter 9).

The second strand concerns those evolutionary changes that have occurred since our line of descent diverged from that of our closest living relatives, the chimpanzee and bonobo. Changes in our biological potential since that last common ancestor can only be inferred by careful archaeology, plus a strong knowledge of modern humans and of the last common ancestor. (Remember that this animal cannot be assumed to be "much like a chimpanzee," as if evolution conveniently left living fossils of our past incarnations lying around to help us understand our origins!) Modern archaeological methods rooted in an understanding of the entire paleo-environment give a firmer basis for interpretation than in the past, but behavior leaves so little material trace after millions of years that even today deductions often have to be revised in the light of some further fragment of evidence.

The third strand concerns the adaptations that endowed the last common ancestor of human and chimpanzee with its particular set of intellectual potentials. These adaptations have their own histories of evolutionary origins: some may have arisen specifically in response to challenges faced by this creature, whereas others will have a more ancient origin. The intellectual capacities of that last common ancestor, and how they arose, can only be discovered by comparative studies of animals.

In the mid-twentieth century the balance among these three contributions—evolution before the human line diverged from that of our closest animal relative, evolution since then, and sociocultural variation—would have been perceived very differently than it is today. Then it was thought that the last common ancestor lived 20 million or 30 million years ago, and that since that time the living apes followed an evolutionary path entirely separate from ourselves. Almost all intellectual advance was thought to have occurred since our line separated from that of any other animal, and to be largely a direct consequence of human language.

In this climate, study of the white rat (or indeed any other animal) was seen as a reasonable model of the nonlinguistic mind, since the building blocks of intelligence in all animals were believed to be much the same. Serious analysis of human intellectual origins and variation was entirely focused on the paleoanthropology of hominids (that is, fossil species more closely related to humans than any other animal) and on cultural studies. Today we know that humans are part of the African great-ape taxon; chimpanzees and bonobos are more closely related to humans than they are even to gorillas, and orangutans are only fairly distant cousins. The last common ancestor of humans, bonobos, and chimpanzees lived far more recently than 20 million years ago. The best current estimate is 4.5 million years ago, about the time of *Ardipithecus ramidus*, a fossil species from northeast Africa thought to be the earliest hominid. This revised thinking makes it much less likely that all human intellectual development could have been packed into such a short span of evolution.

Moreover, modern comparative studies of animals suggest that there are, after all, significant differences in intellectual abilities: it is these differences that give evidence of earlier phases in human ancestry and allow the first glimmerings of the human mind to be studied scientifically. Over this same period, equally dramatic changes have occurred in the way the mind is understood and studied (we sometimes refer to it as the cognitive revolution), with the result that psychology now analyzes mental phenomena as information-processing procedures, adapted by evolution to serve the needs of an organism surviving in a complex world.[1] The influence of this revolution has yet to be fully felt in animal studies, but already our understanding of the evolution of the mind has begun to be couched in cognitive terms.

Phylogenies of Intelligence

The method that is used to interpret comparative evidence and so to reconstruct the ancient mind can best be called *evolutionary reconstruction*. An analogy may help to explain it. Imagine a medical researcher, interested in a genetically caused disease. Suppose she has discovered a population of people more or less closely related to one another—perhaps an island population—in which some individuals have a gene abnormality and others do not. How and when might this have originated? Starting with a family tree going back some generations, she checks with all the living individuals as to whether they carry the defect or not. Plotting the affirmative answers on the family tree, she then traces "up" the tree to find the first branch-point from which all living members with the abnormality are descended. That individual is then the best bet as the ancestor who first suffered the genetic mutation, and when. The distribution of the characteristic in the living members is used to reconstruct its origin in the past. In the same way, we can uncover information about the earliest phases of human behavioral evolution without ever digging up a fossil. The equivalent of a family tree is here an evolutionary taxonomy or phylogeny of our primate relatives.

For each branch-point a common ancestor is implied, from which all the living species on the branch descended (a group of species descended from a single ancestor population is called a *clade*, and an evolutionary taxonomy is composed entirely of a branching tree of clades). Deducing the existence of these ancestors is a reliable procedure, regardless of whether their fossils are ever found, whereas attributing a set of bones to a human ancestor is risky business—that fossil species may have gone extinct and left no descendants. Phylogenies nowadays are usually based on the pattern of shared molecular similarities between species, most often similarities in DNA sequence. Overall change seems to occur at a constant rate over time (somewhat surprisingly, since some parts of the genome are presumably under more active selection than others), so that within limits we can calculate the dates at which these deduced ancestors lived. However, this process depends on calibrating the relative sizes of differences between species against fossil evidence, which itself may not be very reliably dated.

The existence of extinct species, ancestors of both humans and various living nonhuman primates, is therefore deduced from chemistry. Once that has been done, we can go on to reconstruct their behavior and mentality. This second phase involves using "living evidence" from their modern descendants—the nonhuman primates and ourselves. For instance, if an evolved trait of human psychology was found in chimpanzees but *not* in gorillas or orangutans, we could deduce that it had a recent origin, between 4.5 million and 6 million years ago. If orangutans and gorillas also shared it but not monkeys, we would know that it had evolved earlier, before 12 million though after 25 million years ago. For a trait that was more widespread in primates, found in many monkeys as well as in the great apes, an earlier origin still would be deduced, more than 25 million or 30 million years ago. If no living primates share a trait with ourselves—as is true of speech, for instance—then it must have evolved sometime since our ancestral line split off about 4.5 million years ago. (A trait that is found in a scatter of species, none closely related to the other, would most likely have evolved independently in each—so-called *convergent evolution*. A famous example is the wings of birds, insects, bats, and pterosaurs. In the case of intelligence, the large brains of carnivores and cetaceans are evidently convergent with those of primates, since the three groups are not one another's nearest relatives. Evidence of these species' behavior tells us nothing directly about the descent of human intelligence, but may be helpful in detecting the biological functions for which large brains are useful.)

Comparison among the living primates is therefore central to discovering the origins and time frame of modern human behavior: we can separate recent from earlier mental adaptations, and for the earlier ones we can work out when and in what sort of animal the behavior was first seen. In this process it is essential that a broad range of modern species be examined. The discredited "progressive" view of human evolution used a few convenient animal species as models of extinct human ancestors. The enterprise was doomed: there is no valid reason why any particular living animal should resemble an early stage of human ancestry.

Where does the behavioral evidence come from? It can emerge from the laboratory, with the advantage that variables can be individually manipulated, but the trouble is that very few species are kept to test or observe. Or it can derive from the field. Thanks to the careful field studies conducted since the 1950s, we now know a great deal about primates in the wild. Still, under field conditions it is hard to perform accurate experiments, and many of the data must be purely observational. The best approach is a combination of both: reliable recording of naturally-observed behavior in a wide range of species, then a filling of crucial gaps by highly structured observation and experimental testing. (Often experiments will necessarily be on just a few species in captivity.)

Complex Behavior Built on Simple Knowledge

Most animals interact with one other animal at a time, but in monkeys and apes third parties often affect what happens. Social support is therefore very important when individuals compete: where many animals use their strength or weaponry, monkeys and apes rely far more on alliances with other group members to give them power and influence. These alliances are often among close kin, but friendships of nonkin are also found—often lasting over a number of years and giving benefits to both parties. Sometimes the benefits are different for each participant, as in a barter system. Probably the most important trade currency used to build up alliances is social grooming, repaid by support in fights or tolerance at a prized feeding site (which is why monkeys and apes groom one another far more than would be necessary for health purposes alone). Grooming is no doubt pleasant to receive, but it is also time invested in the other, an evidence of commitment that cannot be faked. When major alliances are threatened by minor conflicts, opponents will reconcile afterward, going out of their way to be affiliative with each other. Monkeys and apes lead lives that are socially complex, with a web of influence and obligation that is not immediately obvious. All the same, those factors determine most outcomes that are important to individuals.

Primatologists use their knowledge of individual animals' kinship, friendship, and social rank to understand animal societies. We have learned that monkeys and apes also analyze one another this way. For instance, monkeys who have been attacked by dominant animals sometimes "redirect" their own aggression and threaten weaker, innocent parties—specifically, they tend to attack the young relatives or subordinate friends of their tormentor. In this monkey vendetta, the choice of victims shows that they are well aware of who are the relatives or friends of their opponents. Playing back prerecorded calls to monkeys to study their reactions under controlled conditions has shown that monkeys also know about kinship among other monkeys, the dominance relations between other group members (not just in relation to themselves), and even the membership of groups of which they have never been a part. Monkeys and apes are more than pawns in the web of social complexity; they are socially knowledgeable.

Monkeys and apes rely on this social knowledge in order to learn manipulative tactics, like the use of deception to get what they want. For instance (see Figure 6.1), a female gorilla, living in a small group with a powerful male who prohibits her sexual contact with other subordinate males, may use a number of tactics to give her the freedom she desires. She may just "get left behind" so that she is out of sight of her leader male before she socializes or copulates; or she may invite the male of her choice to follow her, then carry out her actions with unusual quietness, for instance suppressing the copulation calls that she would normally make.

Sometimes these tactics look very smart indeed. Andrew Whiten and I noticed a young baboon several times use the trick of screaming as if he had been hurt, just as he came across an adult in possession of a valued food resource. His mother ran to the scene and chased off the "aggressor," with the result that the youngster got the food. He used this tactic only when his mother was out of sight, and only against targets of lower rank than his mother. Tricks are quite rare—crying "Wolf!" too often simply doesn't work—so researchers tend merely to write them down in their notebooks as interesting anecdotes and not publish them. To get around this difficulty, Whiten and I decided to survey a large number of experienced primatologists and collate their observations of deceptive tactics in primates, looking for recurrent patterns. We found that, although the precise tactics varied, all groups of monkeys and apes occasionally used deception—between species and between individuals. In contrast, the strepsirhine primates (lemurs and lorises) gave no firm evidence of using deception at all. Among monkeys and apes, the frequency of using deception varied, partly because some species were watched far more than others. Even when we corrected for that factor by taking account of the number of studies, some species showed significantly more use of deception than one would expect if all monkeys and apes were equal in intelligence.

There is therefore strong evidence that many species of monkeys and apes show a level of social sophistication and skill unusual for a mammal. Upon what is this social aptitude based? The simplest answer is rather un-

Figure 6.1 Tactical deception in gorilla mating competition. A. An adult female, Pandora, solicits a young male, Titus, with sideways, head-flagging actions. The couple drop behind the group and mate out of sight of the others. B. They are discovered in the act by the leader male of the group, Beetsme (note that the eyes of both Pandora and Titus are directed to the left of the camera position, at Beetsme). Pandora is immediately attacked by Beetsme, but on subsequent days she persists in "illicit" matings with Titus. (Photos by Richard Byrne.)

romantic: it may rely on no more than genetically encoded rules and rapid learning in social contexts. Suppose monkeys and apes are quicker than most other mammals at remembering socially relevant information about conspecifics—information such as rank, kinship, whether they have used a certain tactic on a particular victim recently, whether it worked, and so on. This thesis is consistent with primates' known rapidity of learning in laboratory tasks. They will therefore have a greatly augmented database of relevant facts about their social companions. Much of the process of choosing allies, remembering grievances, retaliating and choosing secondary victims, building and repairing relationships, might then be based on rather general principles—species-typical principles that develop under tight genetic guidance.

Other social manipulation, for instance using deceptive tactics, is more idiosyncratic and certainly depends on learning. However, if they have an ability to make rapid connections between social facts and environmental circumstances, monkeys and apes will tend to develop more elaborate and complicated tactics that give the impression of real understanding even though acquired by trial-and-error learning. For instance, if a young baboon were attacked for approaching an adult with food too close, it would scream in fear. Its mother would rush to its defense—and coincidentally the youngster might end up with the food, a windfall reward.

Coincidences like this are not likely to be common, but if the baboon knows the ranks of individuals relative to its own mother, if it takes note of its mother's presence or absence, and if it can learn from just one natural coincidence, then it is not hard to see how it could acquire the tactic described above. This claim may seem surprising, but learning in a social environment is helped enormously by the tendency of animals to be attracted to the sight of another individual getting a reward. Once attention is drawn to the right place, then trial-and-error learning can be much more efficient.

The ability of monkeys and apes to learn social information very rapidly seems to be closely linked to the size of the neocortex. Among primates, differences in brain size are primarily a function of the enlarged neocortex of some species. The ratio of neocortex to rest-of-brain volume has therefore been used as a simple measure of the extent of selection for larger brains. For monkeys and apes, Robin Dunbar has shown that this ratio varies closely with the size of a typical group (a simple measure of social complexity), but not with environmental factors such as range or day's journey length. (More recent analyses have shown an independent, but much weaker, correlation between increasing neocortex size and a fruit-based diet, possibly reflecting the importance of accurate visual recognition in both fruit selection and social recognition; see Chapter 7.)

Perhaps, then, it was the increasing size and complexity of social groups that served as the major selection pressure promoting brain enlargement in primates. The net result is that monkeys and apes have brains that are on average twice as large as those of a typical mammal of equivalent body size. There is also direct evidence that brain size affects the complexity of behavior shown. To substantiate this, I used a simple index of mental agility, the frequency with which deception is reported for a taxon. After correcting for the number of studies in progress, I found an unambiguous relation with brain size: neocortex ratio predicts how much a species uses deception. The most likely hypothesis at present therefore seems to be that larger brains evolved in response to a need for greater social skill; the increased brain size allowed more rapid learning, underlying the social sophistication shared by all monkeys and apes.

Understanding What You See

It is tempting, when we see an animal use a smart trick to get what it wants, to assume that they understand the situation as we do: to think that the female gorilla mates silently because she *knows* that if her leader male hears her calls he will *realize* that he is being cuckolded; the young baboon screams because he *wants* his mother to *think* he has been hurt. We may be quite wrong. We have noted that the acquisition of deception by rapid learning is sometimes not at all implausible; a deep comprehension of the situation may not be necessary. In fact, experimental evidence indicates that monkeys do not understand the mental states of other individuals (what they want, what they intend to do, what they know).

Captive monkeys respond to a veterinarian in a white coat with a syringe in the same way that wild monkeys respond to a dangerous predator. Using this fact, Dorothy Cheney and Robert Seyfarth performed an experiment on caged monkeys, mothers and daughters. The mothers were allowed to see an approaching "predator" under two different circumstances: (1) when their infants were unable to see what was coming, and (2) when the daughters had just as clear a view as the mothers. The mothers did not seem to understand the difference: they were exactly as likely to give an alarm call when the infants could perfectly well see the danger as when they could not. This inability to understand another's mental state helps explain the puzzle that animals manipulated by deception often do not seem to understand what is happening: frequently, monkeys do not seem to realize they have been deceived and show no resentment, even when it is extremely obvious to human observers.

In contrast, chimpanzees and other great apes give some evidence of understanding the knowledge and intentions of other individuals. Sarah Boysen has employed the same "lurking veterinarian" design. She used pairs of chimpanzees who had spontaneously formed close friendships. One of them was put in a position to see that its close friend was under threat, again from an approaching person with a dart gun. With chimpanzees the results were quite different. The individual who could spot the approaching danger gave intense alarms if it recognized that its friend was unable to see the approaching risk, and indeed the friend took action to avoid the invisible "predator." If both were in a position to see the threat, hardly any alarm calls were given. Chimpanzees seem able to distinguish the mental state of ignorance, and make the connection between a restricted view of the world and likely ignorance of what is happening. This capability is consistent with the fact that in the wild, only chimpanzees—and none of the many species of well-studied monkeys—have ever been reported to teach their offspring in a way that shows sensitivity to the youngsters' ignorance.

Another aspect of understanding mental states concerns deliberate intentions. Josep Call and Michael Tomasello used chimpanzees, orangutans, and children in an experiment where in each trial a researcher would mark one of four boxes by placing a conspicuous object on top; this box alone contained the reward. Once the subjects' behavior showed that they understood this rule, the researcher began to mark *two* boxes in the same trial for the same period of time. He placed one marker deliberately on top of a box and a little later carefully removed it; whereas he contrived that the other marker fell on "accidentally," and when he "noticed," he took it off with slight annoyance. The subjects, for all three species, did not hesitate: they immediately picked the box marked intentionally and avoided the accidental cue. Over time, success at this task decreased for apes and children alike, presumably because the oddity of "accidents" happening every trial began to strike them.

This behavior is consistent with an anecdote of tactical deception from the field. Frans Plooij once deliberately tricked a young chimpanzee, using a "look behind you" gag in order to make her go away. When she returned, she showed by her behavior what in humans we would have to call annoyance; she hit him and then ignored him for the rest of the day, implying an understanding that the trick was deliberate.

More generally, in the survey that Whiten and I carried out, of more than 200 reports of deception in monkeys and apes it was consistently species of great apes (gorillas, bonobos, orangutans, and chimpanzees) that gave qualitative evidence of understanding deception. This evidence was generally a matter of the extreme implausibility of the circumstances that would have to be imagined in order to account for a tactic's acquisition by learning from natural coincidences. Numerically, cercopithecine monkeys used deceptive tactics more frequently than did great apes, yet it was always the apes whose tactics seemed to imply awareness of what they were doing.

Discovering that great apes have some understanding of mental states calls for a reinterpretation of their everyday social maneuvering. For an individual who can realize that others may be withholding useful information or who wants them to switch their allegiances, or who can understand how another would act if it were caused to believe something false, the social world is a very complex place.

Frans de Waal deliberately used the title *Chimpanzee Politics* for his account of the Byzantine complexity of maneuvering among a group of chimpanzees that he had studied. He showed in case after case that a "rich interpretation" of these animals' behavior, one that allowed them an understanding of motive and intention, gave a satisfying and compelling account. As one example, a male who did not have the qualities to become top ranked himself used deft switches of allegiance to gain more effective power than held by either of the two males he supported. Each was only able to gain top rank with his support, and once he had done so and the two began to consolidate their position, he defected to the other. Toshisada Nishida reported similar "king-making" maneuvers in the field. The result was that the supporter gained more matings with females than either top-ranking male.

De Waal found that males who required the support of a powerful ally to hold top rank were always vulnerable, and in the long run a male who gradually built up a broad base of support among weaker males and females was able to hold tenure far longer. He drew a direct and disturbing parallel with the advice Niccolò Machiavelli gave in 1532: "He who attains the principality with the aid of the nobility maintains it with more difficulty than he who becomes prince with the assistance of the common people, for he finds himself a prince amidst many who feel themselves to be his equals." Is this just a parallel, based on similarity of functional power relationships, or an identity, based on similar mental processes in chimpanzee and human? When *Chimpanzee Politics* was published in 1982, even making such suggestive comparisons was considered risky and potentially anthropomorphic; with what we now know of mental-state understanding in great apes, it seems merely ahead of its time.

An apparently rather direct approach to other species' awareness is to present one of them with a mirror: if that individual has a concept of self, it is argued, he or she will be able to discern the identity of the individual "in the mirror." (Of course, if they have never seen a mirror before, most animals—and even most people—are likely to be fooled into displaying social responses to it, such as fear at the sudden appearance of a strange conspecific.)

Gordon Gallup noticed that chimpanzees, but not monkeys, gradually began to show behaviors that implied they had realized the image was their own: looking in the mirror to examine hidden areas of their bodies such as teeth and gums, or making repeated, exaggerated gestures while watching their reflections. To check, he marked individuals while under anesthetic with odorless colored paint, on a part of the body they could see easily and also on a hidden area such as the forehead. After recovery from anesthetic, both rhesus monkeys and chimpanzees noticed and touched the obvious marks, but not the hidden ones. Then Gallup reintroduced a mirror, to which the animals had already become habituated before marking. The chimpanzees, once they caught sight of their reflection, immediately reached up and touched the mark; monkeys showed no such sign of recognizing that it was their own face in the mirror.

Since then, this test has been applied to a large number of individuals among a whole range of primates, and the results are simple: most chimpanzees, bonobos, orangutans, and some gorillas show mirror self-recognition, whereas monkeys do not. Sometimes the mark test or some variant of it has even been given to individuals who do not first show behavioral signs of self-recognition. The results then (which include occasional monkeys mark-touching) are harder to interpret. The mark test is useful only as the final step in a line of evidence that an individual recognizes itself, not as a foolproof entry into the awareness club. Indeed, spontaneous signs of self-recognition may be more convincing in species such as dolphins, where the mark test can only be applied in a contrived way.

Gallup has emphasized that it is only those species that demonstrate other, independent signs of understanding mentality and displaying empathy that show mirror self-recognition. He argues that only individuals with a concept of "self" could ever determine the owner of the face in the mirror. However, knowing that a hidden part of one's body is in some way "the same" as more visible parts like hands and legs is in essence a matter of understanding the physics of mirrors, not the philosophy of mind, and other researchers are less impressed with the ability. Interestingly, James Anderson has found that monkeys are able to use a mirror to see around corners and reach to hidden objects—yet they still show no signs of self-awareness. The jury remains out on the exact significance of mirror self-recognition and whether it shows more general self-awareness.

The Manufacture and Use of Tools

Whether mirror self-recognition is a matter of understanding physical properties or mental states, a comprehension of how physical systems work can certainly lead to the use of an object as a tool to tackle problems that otherwise are hard to solve. Animal tool use has long fascinated psychologists, and pioneering experiments in the early decades of the twentieth century by Wolfgang Köhler in Europe and Robert Yerkes in America frequently set problems requiring tool use to great apes. Most dramatically, one of Köhler's chimpanzees spontaneously fitted two short sticks together to make a pole long enough to reach the food reward that was tantalizing her. Although this work has been a standard citation for "insight" in an animal, subsequent research showed that past experience with objects was crucial. Generally, only chimpanzees allowed to play with sticks during their early development showed insightful tool-making solutions when tested.

Despite the impressive abilities of captive apes, and a scatter of interesting examples of tool *use* across the animal kingdom, tool-*making* was long taken as a hallmark of humanity. For many years the standard popular work on paleoarchaeology was Kenneth Oakley's *Man the Tool Maker*. The scientific community was therefore shaken when in 1963 Jane Goodall first reported tool-making by the chimpanzees of Gombe: as Louis Leakey put it at the time, it meant redefining tool, redefining man, or redefining chimpanzee! Cynics might say that man was therefore quickly redefined as "the syntax user" to preserve our distance from brute beasts.

Today we know that chimpanzees make and use tools for many purposes, but by far the most important role of tool using in the ecology of the species is to gain access to otherwise inaccessible food. Might it be, therefore, that it is not so much chimpanzees' abilities that are special, but only their need to exploit embedded food? In captivity, capuchin monkeys also readily learn tasks that require tools, even sometimes modifying the tool before use. However, a careful comparison shows that this similarity is superficial, and that chimpanzee tool use in the wild includes a remarkable set of abilities, beyond tool use per se.

For instance, it is true that laboratory capuchins and wild chimpanzees both probe for food. But the capuchin is typically presented with a highly simplified world: an obvious hole, food visible within it, and a choice of only a few movable items, one of which is usually an adequate probe. The wild chimpanzee is "presented" with a termite mound, initially with no hole, in the African bush. (To be precise, only a few dozen rather inconspicuous termite nests are located in the home range of 15–20 square kilometers of forest and bush. A random day's walk would miss all of them.) The range of objects with which a mound might be probed (or prodded, or struck) is huge; very few of these natural objects are tried. Even the hole has to be made, and this can only be done in certain parts of the mound where termites are getting ready to emerge, at a certain time of year (the emergence tubes are then capped by the termites with softer mud). The edible termites are not visible at all. Probing must be done delicately, and often the other hand is used as a guide and support. Hasty poking destroys the tool; hasty pulling out dislodges any attached termites. There is, in short, a lot to learn: termite fishing is a sophisticated skill.

Furthermore, termite fishing is not the only elaborated skill that wild chimpanzees show. For instance, in West Africa chimpanzees use two rocks in conjunction as hammer and anvil to break very hard nuts, and in East Africa safari ants are eaten with a technique that relies on bimanual coordination as well as tool manufacture. No Old World monkeys regularly make or use tools in the wild.

Wild chimpanzees sometimes make advance preparations which show us that their tool using is driven by a plan in their minds, not triggered by immediate stimuli in the environment. When no suitable plant is growing near a termite mound, chimpanzees make the tool in advance and carry it to the site. Where there is no suitable large, rounded rock to use as a hammer, they will often pick up and carry a hammer-stone before they get to a tree with nuts to crack. Moreover, William McGrew has shown that the elaborate methods used by chimpanzees vary between sites, in ways that do not relate to local environmental differences: there are local traditions in chimpanzee tool manufacture, just as there are in different human cultures (see Chapter 9). In some populations, termite probes are discarded when the ends become frayed or muddy; in others, the tool is rotated and the other end is used; in yet others, the used end is "resharpened" by biting off the damaged part. The persistence of *inefficient* methods in a population shows that the traditions are passed on socially, by imitation; trial-and-error learning works by gradually homing in on the optimal method, so it cannot readily account for persistent, inefficient traditions.

Food Processing by Gorillas

Thus the traditions of tool use in wild chimpanzees are very different from the innovative tool-using performances of captive capuchin monkeys, and suggest humanlike planning and cultural learning abilities. However, the ability to carry out an elaborate series of actions according to a structured plan or mental program is not restricted to the chimpanzee: remarkably similar abilities (less the use of tools) have been discovered in the gorilla, often stereotyped as a rather dull cousin of the clever chimpanzee. While the lowland or Western gorilla probably has a fruit-dominated diet rather like a chimpanzee's, the eastern populations (which may well constitute a different species) live in more mountainous areas where little fruit is available. This pattern reaches an extreme in the mountain gorillas of Rwanda, originally studied by Dian Fossey.

Mountain gorillas are sometimes portrayed as living in a salad bowl of plants at which they merely nibble when hungry. Indeed, the widespread herb vegetation in their mountain habitat is rich in nutritious plants, few of which are defended chemically. The plants forming the bulk of gorilla diet *are* defended physically: by spines, hard outer casings, stings, or other mechanisms that make eating a challenge (see Figure 6.2).

Jennifer Byrne and I found that gorillas use elaborate techniques to deal with these problems. For instance, when a gorilla eats nettle, he or she quickly amasses a bundle of leaf blades by repeatedly stripping a growing stem with a half-closed hand, removes the stems or petioles that have the worst stings by tearing or twisting them off, then carefully folds the bundle so that only a single leaf underside is exposed when the parcel is popped into the mouth. This procedure cuts down the number of stings that contact the palm, the fingers, and especially the lips. In fact, the complete process involves even more steps, and can be best appreciated with a flow chart (see Figure 6.3). For success, multiple stages have to be sequenced

Figure 6.2 An adult female gorilla eats *Peucedanum linderi*, an important food whose challenge lies in obtaining the soft, edible pith inside the hard casing and manipulating the unwieldy stems, which may be 6 meters long. Note the tiny baby on the female's chest, less than 12 hours old: he will have little difficulty realizing that *Peucedanum* is edible, for he is showered with it daily. (Photo by Richard Byrne.)

correctly; several require precise coordination, with the two hands used together in complementary roles. Most important, some parts of the process
can be repeated until a criterion is reached ("enough food for a handful");
this means that portions of the process are treated as subroutines, building
a hierarchical organization of behavior. These subroutines can be single actions or quite long sequences of actions. We found that the techniques for

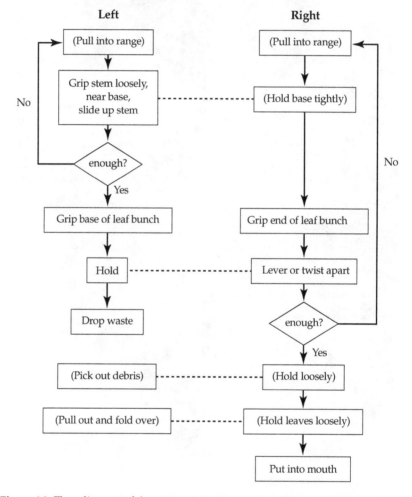

Figure 6.3 Flow diagram of the processing sequences used by a "right-handed"
mountain gorilla in preparing bundles of nettle leaves to eat. The process starts at
the top and ends when a handful is put into the mouth. Actions are shown in square
boxes; for most there is a strong hand preference, and left and right sides of the figure illustrate these preferences. Horizontal dotted lines show when the actions of the
two hands must be accurately coordinated.

other plants, as well as the specific actions involved, vary in logical structure but are equally complex.

Ape Cognitive Superiority

Gorillas are therefore able to build up organized programs of actions, just like the recipes we use to prepare and cook food, and just like the programs of chimpanzees when they set out to use a termite probe or a nutcracking hammer. In humans, behavior such as this is "planned" or "thought out," and these data may be the closest to thinking that we can detect in nonverbal animals. Carel van Schaik has found that members of one population of orangutans make and use tools in ways as elaborate and delicate as those of chimpanzees. If every other great ape shows these abilities, it is likely that bonobos are no different. On the other hand, monkeys dealing with plant problems very similar to those of the gorillas have already been studied, and no sign of imposed, planned organization has yet emerged. Instead, the rather chaotic sequences of monkey actions appear stimulus driven and opportunistic. A monkey/ape difference is thus likely in the complexity of behavioral organization that can be learned without human intervention.

Might great apes have more efficient mechanisms with which to acquire novel skills? To learn complex skills by observation, imitation at an organizational level would be useful: copying the logical organization, including the sequence of actions, hand coordination, and subroutine structure. I call this program-level imitation, and Anne Russon and I have argued that it is within the competence of all great apes. Our evidence is of several different sorts: in the case of mountain gorillas, population consistency in overall technique, combined with great idiosyncratic variation at the level of detail; in the case of chimpanzees, interpopulation differences in tool-using tradition that cannot be explained by ecology; in the case of orangutans, copying elaborate human traditions of skilled action, as shown by Russon's own work. She has observed orangutans in the process of being "rehabilitated" to the wild after illegal capture, and some of them enjoyed copying what humans were doing. One orangutan was noticed following a camp assistant who was tidying paths in the forest, cutting overhanging weeds with a machete, and piling them into heaps that formed a row. The orangutan was also cutting weeds, though with a stick, and making her own row of neat little piles! Another persistently attempted to start a fire, employing the techniques of blowing and fanning glowing embers with a saucepan lid held vertically, decanting kerosene into a jug and tipping it onto a nascent fire, and putting glowing embers next to wood in the hearth. (While watching this process, Russon was unaware that the liquid was kerosene!)

The underlying cognitive superiority of great apes over monkeys might best be described as a difference in forming and manipulating mental representations—data structures in the mind that correspond directly to objects and events in the world. In the case of great apes, there is evidence of mental representation of both complex instrumental behavior and mental states. Manipulation of these data structures in the mind amounts to mental simulation of the world, or planning in everyday terms. In comparison, there is no current evidence that monkeys can acquire complex novel behavior by observation, or comprehend the intentions that lie behind such behavior. The implication is that the great-ape common ancestor would have had a cognitive system qualitatively more like that of humans: able to undertake simple planning, to use others' behavior as a source of new ideas for its own, and sometimes to understand the purpose of others' actions.

Unlike the greater mental agility that all monkeys and apes have than other animals, the deeper understanding of great apes does not seem to be a matter of further neocortical enlargement. Some monkeys have relatively more neocortex of the brain than do some great apes. While it seems that neocortex size affects learning speed, it may be that something else underlies the great apes' differing mentality.

One exciting possibility is that there might have been an organizational or "software" change in great-ape ancestry. A more mundane alternative is that, rather than relative neocortical enlargement, absolute size is what matters. When the total brain volume reaches a certain point, the ability to perceive and organize the structure of instrumental behavior emerges. All the great apes do have brains larger in toto than any monkey or most other mammals. One way of testing our hypothesis is to study the cognition of toothed whales, some of which have brains even larger than great apes. It is difficult to do so, of course, but if whales prove to have minds more like those of apes than like those of monkeys, it would support the absolute-size hypothesis. Already, dolphins have been reported to show a number of suspiciously "apelike" traits: using a sponge as a tool to flush out prey from rock crevices; bringing back injured prey for their offspring to chase and thereby acquire hunting techniques; exploring marks on their bodies with the aid of a mirror; and so on.

Tracing Back along the Roots

We can now use the method of evolutionary reconstruction to impose a time scale on the mental changes that these primate findings imply for remote human ancestry. First, consider the origins of enhanced learning ability, based on relatively enlarged neocortex and giving rise to social complexity (in Figure 6.4, the distribution of these traits in modern primates is indicated by shaded boxes, either hatched or black). Tracing back, we can see that it was in one of the common ancestors of monkeys and apes that these abilities originated. In contrast, the modern lemurs and lorises appear to lack these enhanced abilities, so our best guess is that the common ancestors of all living primates, which lived about 50 million years ago were no more intelligent or socially sophisticated than other mammals.

Second, consider the understanding of complex instrument use and mental states. From the current distribution of evidence (only the hatched boxes in Figure 6.4) we can deduce that one of the common ancestor species of all the living great apes and humans was the first in which individuals realized that others had viewpoints and knowledge different from their own, and could build up novel sequences of actions. This species lived at least 12 million years ago, but not as long ago as 25 million years, when the Old World monkeys split off from the human line.

It is possible now to paint a somewhat fuller picture of the evolutionary origins of some aspects of intelligence that are precursors of our own abilities. The large brains of humans did not appear suddenly in recent human evolution, but first began to enlarge during shared ancestry with modern monkeys and apes, at some point before 30 million years ago. The monkey and ape line of descent must at that time have been exposed to a strong selection pressure for greater intelligence. To judge from the uses of this intelligence by most of their modern descendants, the selection pressure was most likely for coping with more complex social interactions—perhaps as a direct consequence of needing to live in larger permanent groupings.[2] A second major expansion of relative brain size occurred in rather recent human ancestry, when the genus *Homo* originated, but it may have had a different evolutionary function. One branch of these relatively large brained primates, the great-ape ancestors, acquired some extra skills around or before 12 million years ago: they could take account of another individual's knowledge, and they could organize their future actions into plans, perhaps based on their understanding of the behavior of others. The evolutionary origin of the cognitive differences between monkeys and apes can-

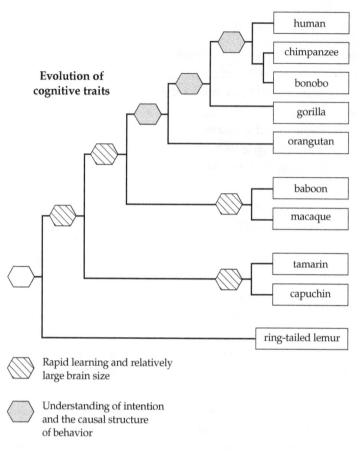

Evolution of cognitive traits

Rapid learning and relatively large brain size

Understanding of intention and the causal structure of behavior

Figure 6.4 An evolutionary taxonomy used to interpret modern data on primate aptitudes in terms of descent from hypothetical ancestor species. Each box on the branching lines of descent represents an ancestor species, known from taxonomic patterns to have existed; the type of shading in the boxes indicates possession of the aptitude, as follows:

Hatched: rapid learning and relatively large brain size. Typical species live in socially complex groups and employ social manipulations that depend on rich and elaborate social knowledge.

Grey: understanding of intention and the causal structure of behavior. Typical species show all of the above traits, but with subtle indications that they have a deeper understanding of social motives and intentions, can learn to recognize themselves in mirrors, and deal with hard-to-process foods by constructing complex, hierarchically embedded programs of actions.

not very well derive from the greater social complexity of ape groupings, since these are not any larger or more complicated than those of monkeys. The sizes of monkey and ape groups overlap completely, and some apes are nearly solitary. Another selection pressure is needed as causal agent in the evolution of great-ape cognition.

We do not know for sure what this second selection pressure might have been.[3] I favor the possibility that the need to procure food efficiently was crucial in the cognitive change, a theme further discussed in Chapters 4 and 5. Great apes are massive animals, heavier than any monkeys; and unlike the quadrupedal monkeys, they are adapted to brachiation, an efficient means of maneuvering in trees but not of long-distance travel. For that, apes are forced to use fistwalking or knucklewalking, at a gait considerably less efficient than quadrupedalism. Wherever they are found, the surviving great apes live in competition with sympatric Old World monkeys. Moreover, the digestion of these monkeys, unlike that of great apes, permits the eating of relatively unripe fruit and rather mature leaves.

Old World monkeys would seem to have held all the cards in the competition with great apes, yet ape species did survive. They must have some telling adaptations to counteract these disadvantages. Their only obvious advantage lies in greater foraging flexibility than monkeys: for instance, chimpanzees are able to gain access to termites inside their rock-hard mounds, while monkeys must wait for the once-a-year evening emergence of winged alates to feast on this prized food source; West African chimpanzees can even eat the meat of *Panda* and *Cola* nuts, far harder than Brazil nuts; one orangutan population can extract grubs and honey from the nests of arboreal bees, whereas the local monkeys cannot; and most strikingly, mountain gorillas have been able to colonize the subalpine and Afro-alpine zones of the volcanoes of Rwanda, where no monkeys can survive. Here their only food competitors are ungulates, whose complex stomachs permit bacterial fermentation to help their nutrient acquisition.

Perhaps evolving an ability to build novel, hierarchically structured programs of behavior enabled better access to resources and more efficient use thereof, allowing some species of ape to survive the intense competition with monkeys. (In fact, the fossil record confirms that many ape species did become extinct over this period.) Whatever happened, it left the survivors with some remarkable skills:

1. A causal understanding of complex behavior, enabling novel schedules of actions to be put together and allowing the behavior of others

to be used as a source of ideas for new action schedules; in other words, nonverbal planning and skill learning by imitation.

2. A degree of understanding of the *intentions* of others—what they want, know, and think; in other words, possession of a *theory of mind*.

Whether these abilities are logically tied together is not at present clear. Still, these very human abilities, which on current evidence date from the period in evolution during which we and the living great apes shared common ancestry, are highly important as a cognitive foundation for development of the unique human capacity of spoken language (see Chapter 8). Evolution of language would be impossible in a species in which individuals could not imagine that other individuals know things that they do not know themselves. The basis of human interpersonal communication is the shared knowledge of the participants, and their understanding of the gaps in individual knowledge. If we did not comprehend this variation, we would not see the point of language.

The mechanism that underlies the productive, generative nature of human language is the ability to build coherent wholes from smaller units in hierarchical fashion—phonemes from distinctive features, words from phonemes, sentences from words. The heart of syntactic structuring is the ability to understand, create, and manipulate flexible hierarchies of motor actions, producing speech. Language is acquired by imitation, but not of the slavish, duplicative variety. Even as children, we do not parrot our parents' words. When a child "imitates" a word, the actual sounds are quite different (a young child does not possess the deep-pitched register of a parent), but the logical structural relations are preserved. This is program-level imitation applied to vocally produced sounds. The structure of another's behavior is comprehended, and so novel structures can be learned, then expressed according to the learner's very different production apparatus. It is intriguing to realize that some of the bedrock of the human language potential dates from 12 million years ago—and may derive from the need of cumbersome apes to get an adequate meal.

7

Brains on Two Legs:
Group Size and the
Evolution of Intelligence

Robin I. M. Dunbar

Group size is a likely measure of social complexity, which in turn may have driven the evolution of intelligence. There is indeed evidence that primates who live in large groups have larger brains. The subsequent evolution of language further promoted effective information exchange and social integration.

L IKE ALL PRIMATES, humans are an intensely social species. Indeed, we probably owe our success as a species to our sociality, since it provides the framework within which culturally transmitted solutions to the problems of everyday survival and reproduction are generated and passed from one generation to the next. Understanding why and how humans are able to engage in such sophisticated social activities will go a long way toward providing an explanation of who we are as a species and where we came from. We are, however, faced with a dilemma when we try to explore the evolutionary origins of behavior. Unlike anatomic components, neither behavior nor the neurocognitive processes that support behavior are readily fossilized. How can we say anything at all about the history of our behavior?

We may attempt to answer this question in three ways. First, we can use a living nonhuman species as an analog for an extinct hominid: whatever the living species does is taken to be a valid model for what the extinct species might have done. Second, we can compare many species of living primates in order to extract general principles that relate some aspect of a species' behavior to other measurable features of its biology; we can then plug in the relevant values for our extinct hominid and read the predicted values for its behavior. Third, we can build systems models of the main components of behavior and interpolate the relevant values for ancestral human populations to see what the models predict. (I shall elaborate on each of these before seeing what they might actually tell us.) A fourth approach that I shall not discuss here is the use of cladistics to identify sets of traits that are common to a set of related species. It is a useful technique, but limited to the identification of behavioral patterns common to an evolutionary clade (for instance, the great-ape clade in the case of fossil hominids).

Three Techniques for Inferring Behavior from the Past

Traditional approaches to reconstructing the behavior of fossil hominids have tended to focus on what might best be called analogical models. A living nonhuman primate (perhaps even a nonprimate mammal) is selected on the grounds that it shares some ecological characteristic with an ancestral hominid population, and it is assumed that whatever this species does is what the hominid species did. Thus, over the years, baboons, monogamous cebids, lions, hunting dogs, and, inevitably, chimpanzees have all been advocated as suitable models for early hominids.

The assumption underlying this approach has always been that two species will exhibit similar patterns of behavior if (1) they are closely related, and/or (2) they share the same ecological niche. Baboons were thought to occupy a similar open-country/woodland niche with (some) early hominids, lions were thought to share a social hunting life style with later hominids, while chimpanzees were thought to be an appropriate analog by virtue of their close taxonomic relationship to modern humans.

This approach assumes that each species has a characteristic way of behaving that is driven by one (or at most a few) key ecological or genetic variables. However, if the last 30 years of research on wild primates have taught us anything, they have taught us that primates are so supremely flexible in their behavior that it is almost meaningless to try to define the "typical" anything for a species. The exemplary fieldwork carried out by Nicholas Davies at Cambridge University has emphasized that even the mating systems of birds can be surprisingly flexible. Obviously, each species' range of possibilities is constrained by its anatomic and neurological structures: no primate flies, for example. Features such as diet (which are heavily constrained by the anatomic design of both the gut and the teeth) are also surprisingly fluid: when pushed to the limit, even the most frugivorous of primate can get by on a diet of leaves—albeit with some difficulty, and only for a limited time.

The short answer is that analogical models do not work; they are often misleading when applied to living nonhuman primates, let alone fossil hominids. A primate species comes into the world with a genetic inheritance that sketches out the broad pathways of its life style, but the details of what it actually does depend on local habitat-specific ecological and demographic conditions. Moreover, even using one species as an analogy for another in respect to one aspect of its biology is no guarantee that everything else the two species do will be the same.

A more promising approach is to search for general relationships that underlie the primate-wide distribution of a given character. Plotting one variable (say, group size) against another (say, body mass) allows us to infer a general relationship that can be described statistically with a regression equation. We can then interpolate the unknown species into this regression equation and read off the predicted value. Body weight is something we can often infer for fossil species because it is related statistically to other more directly measurable features of anatomy (for example, the thickness of the load-bearing limb bones). We can interpolate the estimated body mass of our fossil species into the mass/group-size equation for living species and thereby estimate a value for group size of the extinct species.

This value will, of course, be hedged with (sometimes considerable) error, because every time we interpolate from one equation into another we compound the error associated with each equation. Nonetheless, the mean value predicted by such a procedure is still the maximum likelihood estimate (statistically speaking, our very best guess), and it is certainly better than no guess at all. The more variables we can incorporate into the regression equation, the better: few biological variables are determined by a single factor, so any single-factor equation must make less accurate predictions than one that incorporates many factors. Finally, the predictive power of any such statistical model invariably depends on the degree to which the variables in the regression equation are causally related. Models of this kind will always be more predictive when the relationship they describe is a real cause-and-effect model.

The third approach to inferences about the behavior of fossil species uses multivariate regression equations to build functional models of primate behavioral ecology. These equations differ from those described above in two important respects. First, the individual equations are multivariate and thus are more likely to reflect the underlying complexity of real biological phenomena. Second, the equations model the behavior of individual populations (or even groups) rather than an entire species and thus reflect biological reality.

Behavior, ecology, demography, and the key environmental variables are linked by a structured system of functional equations that reflect the causal relationships between them. If we understand the causal relationships well enough, we can begin to set the equations into a multivariate model that can be used to simulate the behavior of individual populations or even individual groups. These models will always be more powerful than the species-level models described above because they focus on the behavioral biology of individual populations. By the same token, they require much more detailed data for a large number of variables and are therefore more difficult to construct. They may also be taxon specific: a model developed for baboons may apply only to baboons because it reflects the way baboon dietary adaptations (at the physiological as well as the anatomic level) respond to specific environmental variables.

We can ask a number of different questions with this third kind of model, including (1) whether a fossil species could have survived in a given habitat if it adopted a specific kind of ecological niche or body size, (2) why it went extinct at a particular site where we know it went extinct at a particular time, (3) why it was forced to adopt a new ecological niche at a specific point in time, and so on. For each of these questions we may be able to offer quite precise quantitative answers, something that we have hitherto been unable to do with the kinds of analogical models most palaeoanthropologists have been forced to use for want of anything better.

In the following sections I try to show how we can use these last two approaches to illuminate our understanding of hominid evolution. I do not pretend that we can offer detailed answers at this point. My intention is merely to show the way forward to what now seems a realizable goal.

Coevolution of Brain Size and Group Size

The anthropoid Primates, as a suborder, are characterized by unusually large brains: their brain volumes are typically two to three times larger for their body mass than those of other mammals and birds. In addition, brain size varies considerably across primates in absolute terms. Since these two phenomena have always been viewed as part and parcel of the same overarching problem, the main focus has tended to be the more tractable question of why brain size varies across primates.

A number of hypotheses have been offered. These have included suggestions that (1) their larger brains reflect unique aspects of primate ecology, (2) the energetic saving achieved by larger body size allows females to invest spare energy capacity in growing larger-brained fetuses, and (3) primates need large brains to manage the complexity of their social behavior.[1] The latter strategy (the Machiavellian Intelligence or Social Brain hypothesis) notes that the one characteristic that does distinguish primates from other species—and some primate species from others—is the extent to which they are able to use relatively complex social strategies (see Chapter 6). Coalition formation (that is, cooperative alliances formed for mutual protection) has been identified as one such feature, but more cognitive aspects of behavior (such as tactical deception, in which one animal deceives another) have also been identified.

We can exclude developmental explanations at the outset. Though such explanations must necessarily be true, they fail to explain why brain size should have to change: they merely say whether or not the species has the energetic flexibility to evolve a large brain. Some positive selection pressure is needed to drive brain-size evolution against the gradient imposed by the costs of evolving and maintaining large brains (the so-called Expensive-Tissue hypothesis of Leslie Aiello and Peter Wheeler).

We can test the remaining alternative hypotheses by asking which one best predicts brain-size differences across the primates. To do this we need to find appropriate behavioral indexes for each hypothesis and an appropriate index of brain size. The most appropriate index of brain size in primates is the size of the neocortex (the thin outer layer of neural tissue that is wrapped around the old core of the mammalian brain); it is expansion in this part of the brain that has been almost entirely responsible for changes in brain size among primates. Further, it is in the neocortex that all the "smart" cognitive processes occur (those that we would normally associate with conscious thinking). The various ecological subhypotheses can be indexed by variables such as the proportion of fruit in the diet, the size of the home range or length of day journey, and the foraging style. Quantifying social complexity is more difficult, but one factor we can easily measure is group size. It is crudely related to social complexity (at least within taxonomic families) because the number of relationships that an animal has to keep track of increases exponentially with the number of individuals in the group.

When these various behavioral and ecological indexes are plotted against relative neocortex size for different primate taxa, none of the ecological indexes are predicted at all well when the proper statistical controls are used.[2] The only hypothesis that has stood up to careful scrutiny so far is the suggestion that primates have large brains to support the greater social complexity that characterizes their societies (see Figure 7.1).

Mean group size is, of course, a rough measure of social complexity in the sense intended by the Machiavellian intelligence hypothesis. However, further evidence to support the suggestion that the complexity of behavioral interactions may be the crucial issue is given by the fact that, in primates, relative neocortex size correlates with (1) the size of grooming cliques (the mean number of other individuals with whom an animal grooms at higher-than-average frequency), (2) the extent to which low-ranking males can exploit alternative strategies to gain access to reproductive females, and (3) the relative frequency of tactical deception. These findings imply that the information-processing constraints that ultimately limit animals' abilities to hold a group together may actually have much more to do with the mechanisms by which relationships are established and serviced over time than with the simple numerical question of how many relationships need to be remembered. Nonetheless, the general principle implied by the data in Figure 7.1—the larger the neocortex, the larger the group that can be maintained as a coherent entity—remains fundamental.[3]

With this relationship firmly established, we can use the regression equation relating group size to neocortex size to predict group size for

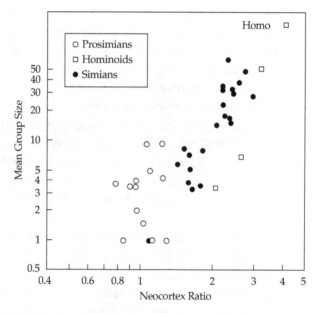

Figure 7.1 Plot of mean group size against neocortex ratio for individual primate genera. Neocortex ratio is defined as neocortex volume divided by volume of the rest of the brain. In most cases, each genus is represented by a single species; in the other cases, the values are averaged across those species of the genus represented in the sample. The value for modern humans is the mean clan size for 21 hunter-gatherer and traditional horticulturalist societies. Note that the hominoids (including humans) lie on a separate parallel grade from the monkeys, who in turn lie on a separate grade from the prosimians. (Redrawn from Dunbar, 1992b, with additional data from Dunbar, 1993, 1998a.)

those species for which we know only neocortex size. When we do this for modern humans, we obtain a predicted group size of around 150, a value that seems surprisingly low when we consider the sizes of modern nation-states. However, we need to bear in mind the definition of a group as used in the nonhuman primate literature: namely, the set of individuals who have strong affiliative relationships with one another, who know and interact with one another on a more or less daily basis, and who maintain some degree of spatial and social coherence through time. In effect, group size for humans is the set of individuals a person knows well and with whom that person has a mutually affective relationship.

Estimates of human group sizes from a variety of ethnographic and sociological sources suggest that there is indeed a characteristic group size of around 125–200 that reappears with surprising frequency in a wide range of contemporary and recent human societies. These groups (villages in both contemporary and Neolithic horticultural societies, hunter-gatherer clans, the number of acquaintances a person has, the basic military unit in most modern armies) all share one crucial characteristic: they consist of a set of individuals who know one another intimately and interact on a regular basis. More important, the sizes of other kinds of characteristic human social groupings lie well outside the 95 percent confidence intervals around the predicted size of 150. In hunter-gatherer societies, for example, foraging groups (usually known as bands or overnight camps) are significantly smaller in size (typically 30–50 individuals), while megabands and tribes (the latter defined as all those who speak the same dialect) are always significantly larger (typically around 500 and 1,000–2,500 individuals, respectively).

Thus there seems to be quite strong evidence that at least one component of human grouping patterns is as much determined by relative neocortex size as are groups of other primates. We have bigger, more complexly organized groups than other species simply because we have a larger onboard computer (the neocortex) to allow us to do the calculations necessary to keep track of and manipulate the ever-changing world of social relationships within which we live.

This being so, we may be able to use data on relative neocortex size to estimate the characteristic group sizes of long-extinct fossil species. In practice, of course, we are unlikely to know neocortex volumes of any fossil species. The best we can expect is an estimate of total brain volume (or at least of intracranial volume, the volume of the inside of the skull). Total brain size is not necessarily the same as neocortex volume, which is what we need to know for our proposed analysis. As it turns out, scaling relationships are available that relate neocortex volume to volume of the brain as a whole, and we can use these to estimate neocortex size of fossil species. (Again, error variance makes the predictions less than perfect.) A more insidious risk is that individual species or genera might have evolved slightly different patterns of brain composition (as did the gorilla).[4]

When we apply these scaling relationships to cranial volume data for fossil hominids and use these to estimate social group size, we obtain the picture shown in Figure 7.2. It plots the group size predicted by the regression equation for Figure 7.1 for every fossil hominid for which we have an

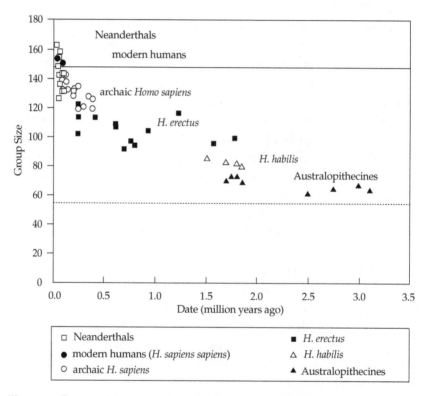

Figure 7.2 Estimated group size for individual fossil hominid populations plotted against the time in which they lived. Group size is determined (from the relationship shown in Figure 7.1) using neocortex ratios estimated from fossil crania. The lower, dashed line indicates the mean community size for living chimpanzees (55); the upper, solid line, that for modern humans (150). A population is defined as all fossil crania obtained from a given site over a 100,000-year period. (Based on Aiello and Dunbar, 1993.)

estimate of cranial volume against its dated time. The main species are shown individually.

What these data suggest is that group sizes among the australopithecines (the very earliest members of the hominid lineage) were fairly stable over time and broadly within the range of those for living chimpanzees. A small (but possibly significant) jump in group size occurs with the appearance of the first members of the genus *Homo,* followed again by a long period of relative stasis. Toward the end of the *Homo erectus* period an exponential rise in group size begins and reaches its apogee not with modern humans *(Homo sapiens sapiens)* but, ironically, with the much maligned Neanderthals *(Homo sapiens neanderthalensis)*—although, to be fair, the difference is relatively small and there is considerable overlap in the distribution of points for the two subspecies.

In summary, this approach seems to be able to provide us with at least a guideline for group size and how it might have changed over the course of hominid evolution. We are now able to say something more than that group size increased over time: we are able to put some reasonably realistic numbers onto the pattern. I shall return to the implications of these results, but first, let me explore the second of the approaches outlined earlier.

Relating Group Size to the Environment

Although a species may have a "typical" group size whose limits are set by its cognitive capacities and neurologically hardwired predispositions, this typical group size simply reflects the long-term average conditions that the species can expect to encounter. With species as long-lived as primates, completely hardwired behavior would be a recipe for evolutionary disaster: environmental conditions can undergo substantial change on a scale much shorter than an animal's life span, and the species as a whole would face considerable difficulty in being able to adapt fast enough at the genetic level to environmental changes on this timescale. Instead, large-bodied, long-lived animals rely on behavioral flexibility to tide them over shorter-term fluctuations in habitat conditions. Primates have been able to take this very general principle to its extremes. We can therefore expect populations to vary in typical group size in ways that reflect the selection pressures acting on the animals on a day-to-day basis.

We come now to our second approach to the behavioral ecology of extinct species. It is based on using comparative data to derive taxon-specific models that describe the overall functional relationships between key behavioral or ecological variables and the environmental or climatic factors that influence them.

Two main factors will influence group size. One is the benefit of living in a group. On the basis of accumulating evidence, we take this to be predation risk for most primates. Defense of feeding and other resources may also play a role, but it seems that in most populations predation risk predominates; only when it is very low will resource defense be sufficiently intrusive to become a driving variable. Primates typically minimize the risks of predation by increasing group size.

Such increase brings the animals into conflict with the second factor, the costs of group living. These take two forms. Ecological costs are created by the impact of group size on feeding and travel time budgets. As group size increases, animals are forced to travel farther in order to find a sufficient number of food patches to satisfy the requirements of all group members. Increased length of day journey will result in more time spent traveling, or a faster rate of travel if time spent moving is to remain constant. In either event, the energetic costs of travel must be met by increased feeding time.

Since the social groups of most primates are bonded (that is, given coherence and stability through time) by social grooming or other forms of social interaction, an increase in group size will also mean that more time has to be devoted to social interaction; otherwise the dispersive forces of ecological competition will result in the group's fragmenting and undergoing fission. An upper limit will thus be set on group size when the habitat-specific requirements for feeding, travel, and rest exhaust the available daylight hours. When this upper limit is low (say, lower than the minimum size required to reduce predation risk to some acceptable level), the species will not be able to survive in its present habitat. This conclusion allows us to determine the limits on the species' ability to occupy habitats successfully.

We have developed systems models that allow us to predict the ecologically maximum group size imposed by time-budgeting costs for three taxa (Papio baboons, gelada, and chimpanzees), and we have been able to derive an equation for the minimum permissible group size (as a function of predation risk) for one of these (the baboons).[5]

The equations that form the core of the model are derived from time-budget and behavioral ecology data for individual populations of a given taxon. They relate these variables to habitat-specific climatic variables that act both as proxies for food resources and their distribution and as direct influences on behavior (for example, via thermoregulation). Subject to the constraints imposed by the cognitive limits on group size discussed above, these equations allow us to define the range of group sizes that a species can successfully support in a particular habitat. Where within this range of group sizes a particular population actually lies will depend in part on whether it places a greater weight on maximizing reproductive rates (in which cases it will opt for the lowest group sizes commensurate with minimizing predation risk) or on maximizing adult survival (in which case it will opt for the largest group sizes commensurate with a balanced time budget).

The baboon model (see Figure 7.3) has been tested in some detail and has been shown: (1) to predict the time budgets of populations not included in the data set from which the model was derived, (2) to predict

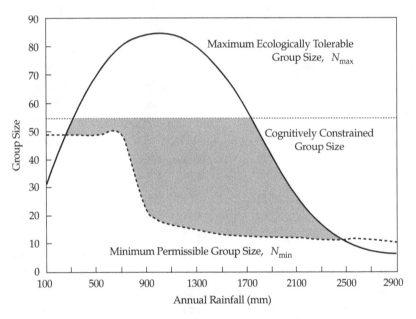

Figure 7.3 Systems model of baboon socioecology, showing the range of group sizes possible for baboons under different rainfall conditions. Temperature is assumed constant at 20 degrees Centigrade (approximately the median value for baboon habitats). The upper, solid curve plots the maximum ecologically tolerable group size (the largest group size that will allow the animals to balance their time budgets). The lower, dashed curve is the minimum permissible group size (the minimum required to reduce predation risk to an acceptable level). The straight dotted line marks the approximate upper limit for stable groups, as determined by neocortex size. Baboons can survive in any habitat where maximum group size exceeds the minimum permissible under predation risk, provided this does not exceed the cognitively constrained group size. (Redrawn from Dunbar, 1996.)

seasonal and interannual changes in time budgets within populations, (3) to predict patterns of fission within populations, and (4) to predict the locations where the species is actually found. The fourth test has also been successfully passed by the gelada and chimpanzee models.

The models seem to be surprisingly robust, especially in light of the fact that the driving variables in the models are just four climatic variables (mean annual temperature, mean annual rainfall, and two measures of rainfall seasonality). One reason is that only three variables are actually needed to describe a habitat's climate. Using principal components analysis, Daisy Williamson has shown that nine very different measures of climate from 218 weather stations throughout sub-Saharan Africa clustered into three discrete sets (essentially annual rainfall, mean temperature, and rainfall seasonality).

Of the four tests listed above, the last has significant implications for our present interests. If we can predict where a species can and cannot occur on the basis of its maximum ecologically tolerable group size, we can thereby explore whether fossil populations could have survived at specific locations. Moreover, since the operational core of the models lies in time budgets and day-journey length, we can use the models to predict these aspects of the behavioral ecology of extinct populations as well.

What makes this enterprise possible is that we can reduce the model to two key climatic variables (rainfall and temperature) because we can estimate the two seasonality variables (albeit with some error) from these two variables. It is usually possible to provide a reasonable estimate of rainfall and temperature for fossil sites; thus, we can look more closely at the ecology and demography of extinct species. The precision with which we can do so will, of course, depend on the extent to which the palaeoclimatologists can give us reasonable estimates for individual fossil sites. Even with the less-than-perfect climatic data currently available, we have been able to explore the behavioral ecology of some of the extinct species of baboons (genera *Parapapio* and *Papio*) and geladas (genus *Theropithecus*) and show that maximum tolerable group sizes were significantly lower at sites (or horizons) where these species did not occur than at those sites where they did occur. Although the analyses do not rule out the presence of these species altogether from locations that have thus far not yielded fossils, they do suggest that, if actually at these sites the species would have found it extremely difficult to maintain a presence there. Improvements in the precision of the models and better estimates of climatic variables should in due course allow us to sharpen these assumptions.

We can perform these analyses only for species that are ecologically identical to the living taxon (genus) for which the model was developed. We know from detailed comparisons of the three models that the equations linking time-budget variables to climatic variables vary between taxa as a consequence of the species' characteristic feeding ecologies. Even though the same variables enter the equations for two different taxa, the slope parameters are quite different.

We can, for example, show that the feeding-time equations for geladas, baboons, and chimpanzees reflect their contrasting ecological niches and their preferred food sources: the gelada time budgets respond to the availability and nutritional content of grasses, the baboons to bush-level vegetation, and the chimpanzees to tree-level vegetation. Each of these vegetation layers responds in a different way to the fundamental climatic variables: grass predominates in cool conditions, bush in hot-dry conditions, and forest in warm-wet conditions. As a result, the models predict only limited ecological (and hence geographic) overlap of these three taxa—as is the case. Since there is no reason to suppose that time budgets alone would predict such district ecological separation between species, this test of the validity of the models is strong. Not only do the models predict what we see, they also provide an explanation of why these geographic patterns emerge.

We can use this result to help us answer a slightly different question about evolutionary history. If we have reason to believe that a fossil species behaved, ecologically, rather like one of our three modern species but later changed its ecological style, we can use the relevant model to ask when, and under what environmental conditions, the species was forced to alter its life style. We know, for example, that the australopithecines were essentially chimpanzeelike but evolved into species (of the genus *Homo*) that were more wide-ranging and carnivorous. We have yet to look at this later transition, but Daisy Williamson has used the chimpanzee model for a preliminary look at the australopithecines. Her results suggest that even the earliest australopithecines must already have been exhibiting dietary and other ecological differences from modern (and presumably contemporary ancestral) chimpanzees: the chimpanzee model suggests that chimpanzees could not have survived at most of the sites where we actually find them. The only explanations for these results are that (1) the fossils accumulated at those sites as a result of external forces (for instance, predators bringing corpses back to their lairs and other taphonomic processes), (2) the species occupied a very narrow microhabitat, or (3) they had already begun to exhibit significant dietary differences from modern (and ancestral) chimpanzees. The fourth possibility, of course, is that the models themselves are flawed, but this is implausible given the demonstrable precision of the models when applied to living populations of the respective taxa. Thus, these models provide evidence that is quite independent of the conventional anatomic evidence, to suggest that australopithecines might not have been ecological chimpanzees.

This approach has considerable promise as a tool for exploring ecological aspects of fossil species with a level of precision far beyond that currently available. Contemporary analyses are limited, by and large, to broad generalizations about ecological niche that do not allow us to make any truly useful statements.

Behavioral Implications

Several consequences at the behavioral level follow from the fact that the hominid lineage evolved large groups during the later stages of its evolutionary history. I consider, first, the bonding of groups and, second, the disruptive forces that threaten the breakdown and dispersal of large groups.

The social groups of Old World monkeys and apes bond mainly by the intensive use of social grooming. Indeed, a more or less linear relationship exists between their group size and the time spent grooming. (It is not so clear that this relationship also applies to New World primates, who typically have much smaller group sizes than those of their Old World cousins.) If we set a regression equation through these data, we can use this relationship to predict the amount of time that hominid groups would have needed to devote to social grooming if they were to bond their social groups in the time-honored way of Old World primates. We find that grooming-time requirements would have been well within the limits for modern apes and Old World monkeys (up to 20 percent of time spent grooming) until well into the *Homo erectus* phase. Toward the end of that period, the social time requirement began to increase very rapidly, reaching a maximum of around 42 percent of the total daytime for modern humans with their group sizes of about 150 (see Figure 7.4).

No species that has to earn a living in the real world can afford to devote so much time to social interaction. It would mean either being unable to forage enough to meet its nutritional requirements or spending less time resting than the habitat demands (with consequences in terms of thermal overload, water stress, and extra feeding time—because when ambient temperatures rise above normal body temperature, energy requirements increase as the body works to prevent overheating). The implication is that the later hominids could not have lived in such large groups, or else they had to find an alternative mechanism for social bonding that made more efficient use of social time than grooming does. One of the limitations on grooming is that it is of necessity a one-on-one activity: several individuals cannot easily be groomed at the same time.

Although we do not yet fully understand how grooming works to bond social groups, one likely mechanism is that it enables animals to spend time in close proximity, thus facilitating the development of a sense of mu-

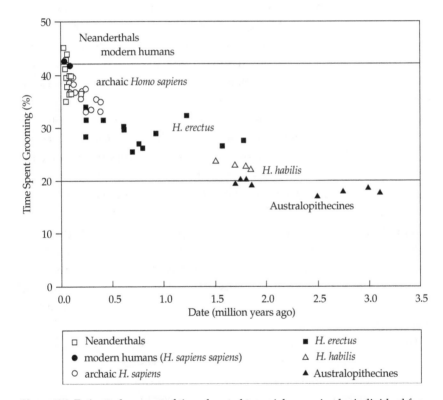

Figure 7.4 Estimated amount of time devoted to social grooming by individual fossil hominid populations, plotted against time when they lived. Grooming time is predicted from the group sizes shown in Figure 7.2, using the regression equation for grooming time against group size for Old World monkeys and apes. The lower line at 20 percent indicates the highest amount of grooming time ever observed in any nonhuman primate. The upper line at 42 percent represents the amount of grooming time required for a group of 150. Modern humans typically spend 20 percent of their waking time in active social interaction. (Based on Aiello and Dunbar, 1993.)

tual trust (and reliability as allies). If this is so, then one way of increasing the size of groups may be to increase the number of individuals with whom one engages in these kinds of social interactions. Language allows living humans to do just that because, unlike grooming, we can converse with several people at the same time. To get from the mean group size for chimpanzees (55) to that for modern humans (150), any such bonding mechanism would have to be about three times (150/55) more efficient than grooming: given that grooming is a one-on-one activity, language ought to be a one-on-three activity. As it turns out, the maximum size of natural human conversation groups (one speaker plus three listeners) is exactly the size required.

The results shown in Figure 7.4 do not necessarily mean that language evolved the moment hominid group size exceeded that which required about 20 percent of the daytime to be devoted to social grooming. Living primates like the gelada already demonstrate that they can use vocalizations to lessen the impact of this effect as group size drifts above its critical limit. In these species, vocal exchanges (for example, of contact calls) seem to act as a form of grooming-at-a-distance so that animals can continue to reinforce a relationship while engaged in other activities (such as feeding or traveling). This kind of time-sharing may be an important buffer for those species with vocal systems that can be used in this way.

All the same, a limit is likely on the size of group that can be maintained in this dual-process fashion. Vocal exchanges are inevitably less efficient as a bonding mechanism than grooming; the physical contact of grooming both releases opiates that act as an immediate reward and (either independently or as a direct consequence) reduces physiological stress. These effects probably help to create the sense of trust and warmth between the partners that seems to be necessary for grooming to work as a bonding mechanism.

Thus, while the early members of genus *Homo* probably did exploit the opportunities provided by vocal exchanges to increase group size, there remained an upper limit beyond which this mechanism would not work. To breach that barrier, something much more efficient was needed—namely, language. It is not clear just where this barrier lies, but a likely guess is about midway between the limiting 20 percent of grooming time seen in primates and the 40 percent required for the groups of 150 exhibited by modern humans. With a barrier at about 30 percent, Figure 7.4 suggests that the evolution of language (in the sense that we now understand it) probably occurred half a million years ago with the first appearance of archaic *Homo sapiens*. Some populations of late *Homo erectus* may also have begun to show this transition (although these presumably would have been populations in the direct ancestral line of modern humans).

Language, of course, has a further benefit over conventional primate bonding mechanisms: the fact that we can use it to exchange information about ourselves and other members of our social network. Primates can only know about what they see for themselves. Should an ally defect or renege on its coalition partner while the partner is not there, that partner will never know and may thus be exploited with impunity. Language provides us with a medium of information exchange that overcomes this limitation. We can monitor what is going on elsewhere within the network, allowing continuous updating of our knowledge of the matrix of relationships. We can also use language to advertise our own value as a mate or ally, and we can use it to police defaulters. ("Never go out with so-and-so; you always end up having to pay the bill.") Although we can (and obviously do) use language for much more sophisticated purposes (instruction, science, literature), studies of the content of naturally occurring conversations suggest that we typically devote 60–70 percent of our conversation time to social topics.

One more relevant fact is that the amount of time spent by humans in social interaction (principally, though not necessarily entirely, in conversation) averages out at exactly 20 percent (based on studies of time budgets carried out in Britain, Nepal, and various parts of Africa). In other words, we spend as much time in social interaction as the upper limit seen in Old World primates—but language allows us to use that time much more efficiently.

8

From Primate
Communication to
Human Language

Charles T. Snowdon

Does animal vocal communication—from the babbling of marmosets to the song-learning of birds—provide clues about language origins? Since many features of human language are present in the communication of other species, meaningful parallels can be drawn. No nonhuman animal shows all of the human features, however.

O F ALL THE TOPICS in this book, the origin of language is one of the most difficult to imagine emerging from our nonhuman primate ancestors. Evidence of cooperative hunting, of the cultural transmission of tool use, of empathy and reconciliation, of manipulating the behavior of social companions is clear in great apes and occasionally in some monkeys. In these areas it is easy to see much of our own behavior reflected in the behavior of apes and monkeys, and vice versa.

But language seems to be fundamentally different. There seems to be nothing approaching the complexity of our vocabulary, our grammar, the concepts and ideas we can express about ourselves and our world, in our knowledge to date of primate communication. This gap creates a potential crisis for an evolutionary biologist: Is language a special creation unique to our species, or can we find some evidence for precursors of language and speech among our nonhuman ancestors?

This question seems to me to raise a false dichotomy. Each species has a uniquely evolved system for communication, so it is not surprising to find species-specific forms of communication. At the same time, we expect to find some commonalties in how signals are formed, transmitted through the environment, and perceived by others—in how divergent species develop common solutions to similar problems in managing social living. So while on the one hand we need to acknowledge the uniqueness of human language, we should not be deterred from trying to find commonalties and continuities with other species. I present here an optimistic view that the study of nonhuman species does hold much promise for understanding the evolution and ontogeny of human language.

To evaluate the issues relating to the role of nonhuman animals in understanding the evolution and development of language, I examine five main points:

1. What are the criteria used by linguists and psychologists to define language? Before we can ask how studies of nonhuman species contribute to the understanding of language origins, we must reach some agreement on definitions.

2. What parallels to language can be found in nonhuman animals? I argue that many precursors of language can be found in nonhuman animals, though no single species displays all of the characteristics of language. Furthermore, the production and perception of speechlike sounds have a long phylogenetic history, suggesting that we have built human speech on ancient biological foundations.

3. Is there a language instinct? One of the strongest arguments for the uniqueness of language has to do with universals that appear across all languages, and the claims of rapid and universal acquisition of language by infants. How strong are these universals? What role do social interactions, teaching, and reward play in language development?

4. How do communication skills develop in nonhuman animals? If developmental processes are found in nonhuman species similar to those found in human infants, and if few universals exist in human language development, then the development processes underlying language may not be uniquely human and can be explained in terms of general principles that apply to many species.

5. How did language evolve from animal communication? The ultimate problem for anyone trying to argue that language has evolved from nonhuman animal communication is to provide a plausible scenario. Can the complexity of language be derived from simpler processes? Can we find potential adaptive value in increasingly complex communication skills?

What Is Language?

Because language appears to be qualitatively so different from what we see in other animals, and perhaps because of our own fascination with language, many popular books and scholarly articles have proposed theories about language origins. Almost every author has a different view of the critical features that define language.

Generative Grammar and Anatomic Uniqueness

More than 40 years ago Noam Chomsky argued that nonhuman species were irrelevant for understanding language. Key to Chomsky's argument was the idea of a generative grammar that allows for the creation of an infinite number of sentences in language. This occurs not only by sequencing several clauses, as in the eight-page-long sentences of some German novels, but also by embedding ideas within sentences. Thus, "He gave the sister that looked after her mother who was ill an extra Christmas gift to thank her," where the main subject and predicate ("He gave the sister . . . an extra Christmas gift") are separated by another clause. Such sentences cannot be understood by simple linear analyses and must require, according to Chomsky, some innate grammar processor. Because children learn to speak and learn the rules of grammar so quickly, Chomsky, and subsequently Steven Pinker, argued for the presence of an innate Language Acquisition Device.

Michael Corballis also sees generative grammar as a unique characteristic of human language, but he extends this generative ability beyond language to include mathematical reasoning, tool-making, music, and art. For Corballis, the specialization of the left hemisphere for language is critical. No other species is as consistently right-handed as humans in fine motor tasks and no other species has such specialized cortical areas for language. In addition, specialized areas within the left hemisphere are involved in specific language processes. The best-known of these is Broca's area, which seems mainly involved in speech production, located near the area of the cortex that controls body movement. Wernicke's area appears to be involved in speech perception and comprehension and is located near the areas of the cortex involved in the perception of sounds. Early studies of nonhuman primates have not found specific cortical areas critical for the production and comprehension of communication sounds (most of the areas were subcortical), but areas homologous to Broca's area and Wernicke's area have been found in nonhuman primates and are important for individual recognition and for controlling movements of facial muscles, tongue, and larynx. Advances in noninvasive neuroimaging methods allow the study of sign- and symbol-trained chimpanzees. The early results suggest that linguistic-like processing may occur in the left hemisphere in apes as well.

Philip Lieberman finds anatomic uniqueness in the structure of the human vocal tract. Our larynx is located significantly lower in the throat than in other species, which allows us to produce a unique set of vowels [I], [a], and [u] as in *see, saw,* and *sue.* According to Lieberman, nonhuman primates, Neanderthal man, and human infants lack this anatomic specialization and thus cannot produce the full range of human speech sounds.

However, later studies challenge Lieberman's conclusions about Neanderthal man.

Symbolic Communication

Many see symbols as the key to human language. Our large vocabulary of words allows us to speak of objects (past, present, and future), to discuss ideas, to create mental images of scenes far away. A word like *dog* can refer to an animal that is present; to dogs previously encountered but not currently present; to pictures, sounds, or smells of dogs; and even as a metaphor for the behavior of dogs as in "She doggedly pursued her degree."

Robbins Burling sees words as critical to human language. He argues that grammar has little value without words. Although he acknowledges that some animals have calls that can be interpreted as symbolizing, say, a particular predator, and that chimpanzees and dolphins can be trained to recognize an arbitrary sound or visual pattern as representing a specific object, Burling argues that no nonhuman animals can use their symbols in the multiple ways that we use words.

For the psychologist William Noble and the archaeologist Iain Davidson, symbols must go beyond words. For them the key event in language is the ability to represent signs symbolically, and for them the origins of modern language extend only from the appearance of cave paintings, approximately 40,000 years ago. Paintings and other depictions imply the ability of organisms to be aware of themselves and their environment and to represent this awareness symbolically. Noble and Davidson argue that the major evolutionary expansion of brain capacity occurred immediately before or coincident with the first expression of art.

Awareness and Conversation

For many linguists and psychologists, self-awareness, consciousness, the ability to reflect on past and future events, and the ability to take the perspective of others—which is essential for pedagogy to be effective—are critical to language evolution. Some, like Pinker and Noble and Davidson, see these higher mental processes as precursors to the emergence of language, whereas the linguist Derek Bickerton argues that development of a symbolic language with a generative grammar is what released early humans to develop awareness, reflexiveness, and the other aspects of our intelligence.

David Premack has argued that conversation is a more important characteristic of language than syntax or words. Conversation requires that we be able to take the perspective of those with whom we converse. Conversation implies rules about turn-taking and reciprocity, and entails not only the sharing of a common code but also shared assumptions about each other. Robin Dunbar (see Chapter 7) takes this idea a step further by noting the importance of grooming in maintaining social bonds in many species of monkeys. With the increased group sizes and resulting social complexity of human social groups, we can no longer maintain appropriate social relationships through mutual grooming, so instead we use gossip as conveyed through conversations to maintain social relationships.

Hockett's Design Features

Linguists have no consensus on what defines language, and it is difficult to translate many of the proposed criteria into hypotheses that can be tested empirically in nonhuman animals. Fortunately, there is another approach. Forty years ago Charles Hockett first proposed a set of design features for language that he said could be used to evaluate the abilities of nonhuman animals. Hockett's framework is one of the few attempts to provide specific and concrete definitions of the features of language. These design features are enumerated in Table 8.1. Some of the features (auditory-vocal channel, broadcast transmission, directional reception, rapid fading, interchangeability, complete feedback, and specialization) are common to most acoustic communication systems. In the following section I evaluate whether nonhuman animals meet any of the remaining design features, and I examine other parallels between human language and nonhuman communication.[1]

Parallels in Nonhuman Animals

To evaluate whether language has its origins in nonhuman species, we need to consider the communication abilities of nonhuman animals. First, I present some different methods for approaching this topic. Next, I return to Hockett's design features to see how well nonhuman animals meet these criteria. Then I explore some parallels in production and perception of speech sounds by humans and examine whether similar processes apply to the perception of their own sounds by nonhuman animals. Finally, I illustrate some parallels in nonverbal communication.

Methods of Studying Nonhuman Animals

The two primary sources of information about the origins of language have been the archaeological record and studies of our nearest relatives, the great apes. Both sources present problems. Only rarely is behavior preserved in the archaeological record (footprints at Laetoli, the Rosetta stone, cuneiform tablets), so conclusions drawn about language are necessarily speculative and often based on assumptions that can never be confirmed or refuted. Great apes are our closest relatives, but humans and apes diverged several million years ago, and each species has been subject to different adaptations over this period. The complexity of vocal communication found in some other primates rarely appears in great apes. In general,

Table 8.1 Hockett's design features of language (adapted from Hockett, 1963)

1. Auditory-vocal channel (but sign languages are accepted as natural languages)
2. Broadcast transmission (sounds carry in all directions)
3. Directional reception (listeners can tell direction of sound source)
4. Rapid fading (sounds decay rapidly, so listeners can process many signals in a sequence)
5. Interchangeability (users produce sounds that they comprehend)
6. Complete feedback (communicators perceive sounds that they produce)
7. Specialization (energetic costs of signals are small)
8. Semanticity (signals are associated with states, events, or objects)
9. Arbitrariness (signals are not iconic or onomatopoeic; signal does not resemble action or event to which it refers)
10. Discreteness (units are individually differentiated and do not intergrade)
11. Tradition (signals can be transmitted to others by teaching or learning)
12. Displacement (referring to objects or events remote in time or space)
13. Openness (forming new utterances through blending and transforming)
14. Syntax (orderly or predictable sequencing of simple elements)
15. Prevarication (conscious lying or deceit)
16. Learnability (speakers of one language can learn other languages)
17. Reflexiveness (communicating about communication)
18. Duality of patterning (combining different sounds into words and words into phrases; with different orders having different meaning)

apes seem relatively quiet compared with other species and thus might not be the best models for studying the evolution of language. To date, most research on great apes has focused on long-distance calls between different groups rather than the potentially more complex signals used for coordinating behavior within social groups.

A third approach complements the other two. Many years ago Peter Marler suggested that songbirds might be better models for studying speech and language, since birds are much more vocal than apes and many monkeys and since birdsong ontogeny shares many parallels with human speech development. By the same reasoning, highly vocal arboreal Old World and New World monkeys might provide important insights. My research on marmosets and tamarins shows them to be highly vocal and to live in small family groups—a social system sharing many similarities with that of most humans.

Beyond the choice of species is the selection of how to study animal communication. Several scientists working with chimpanzees, bonobos, orangutans, dolphins, and parrots have used some analog of language: complex symbols presented on a keyboard, signs from American Sign Language (ASL), artificial vocal or visual signals, or, in the case of parrots and bonobos, spoken language. Several people have used a form of ASL to train great apes. These apes are reported to have acquired a relatively large vocabulary of signs that can be used in sequences of two or three signs. The psychologist Duane Rumbaugh developed a symbolic communication system using arbitrary lexigrams to represent words that could be used on a computer. Several chimpanzees and bonobos have learned to use lexigrams to make requests and to answer questions (see Figure 8.1). In most cases the animals have acquired a large number of distinct lexigrams that can be used according to a simple syntax.

Lewis Herman has trained dolphins with both a whistle-based and a visually based symbolic system. Dolphins can respond to novel sequences of commands given through these systems. Irene Pepperberg has taught parrots to comprehend and produce English words. Her parrots can respond appropriately to questions such as What is this? What is this made of? What color is this? and How many are there? These studies are all very useful in showing us what nonhuman species can do if given adequate training. The animals that turn in the best performances are often, however, unusual members of their species, selected for intense training because of their ability to develop a relationship and work well with human trainers.

Figure 8.1 The chimpanzee Ai, using a computerized keyboard to communicate with her trainer and answer questions. The lexigram appears on the screen at the right, and Ai matches it by pressing the appropriate key from the left side. (Photo by T. Matsuzawa.)

The alternative approach to language training is to study the communication of animals in natural or seminatural habitats without attempting to train specific behaviors. This approach has the virtue of illustrating what animals do spontaneously in the context of normal social activities, but it may not illustrate the capacities a species might demonstrate if given sufficient training. Thus, studies of wild apes may never reveal what the language-trained apes can accomplish as a result of their extensive training. I prefer to study the natural communication of animals, for this approach shows best how communication has evolved in response to natural selection.

Hockett's Design Features in Nonhuman Animals

It is easy to find examples from birds and mammals in which each of Hockett's first eleven design features can be observed. Dorothy Cheney and Robert Seyfarth spent many years studying vervets in East Africa. These small monkeys share the savannas of Africa with many other species, including numerous predators. Cheney and Seyfarth found three types of alarm calls given by vervet monkeys. The calls are obviously *vocal* and are *broadcast* through the environment with a *rapid decay time*. Senders and receivers can use the signals *interchangeably*, so Joe may call today while Sarah responds, and Sarah can call tomorrow and Joe responds. Each type of call is associated with a different type of predator: snake, leopard, or eagle—but the vervet calls do not resemble the calls of the predators *(arbitrariness)* and each type of predator call differs from the other *(discreteness)*. Elegant studies used auditory playbacks of these calls in the absence of predators, and the monkeys reacted to the broadcast calls as if the actual predator were present. Monkeys hearing the leopard alarm call rapidly climbed into trees, while those hearing the eagle alarm call ran out of the trees and took cover in the brush. Thus, the calls have a *semantic* referent, an *arbitrary* structure, and are *discretely* different from one another. We do not know for certain how much young monkeys must learn to produce appropriate call structure, but we do know that vervet monkeys must know at least some of the appropriate usage and response to calls of others; therefore, *tradition* is involved in these alarm calls.

The remaining seven criteria from Hockett are more difficult to document, but some evidence can be found. *Displacement* involves referring to objects or events remote in space and time. The waggle dance of honey bees communicating direction and distance to a food resource is one of the best examples of displacement. In mammals, depositing a scent mark that is "read" hours or days later can provide a record of the type of organism and its movements through space and time. It is curious that so far no evidence of displacement exists in vocal communication.

Openness refers to the ability to form new utterances by blending or transforming utterances, so we humans have the relatively new verbs *to xerox* derived from a brand name for a dry ink reproduction method (xerographic) and *to fax* derived from the noun *facsimile*. Pepperberg noted that when one of her parrots, who had been trained to name pale yellow fruit as bananas and bright red fruit as cherries, used the blended word *banerry* when first exposed to a red apple with its pale yellow interior.

Simple *syntaxes* have been reported in several nonhuman primates. The cotton-top tamarin, a small monkey from Colombia, has two calls used in territory encounters—one typically given by females, the other by males. However, at the peak of aggressive interactions, both calls are given in sequence by both sexes, with the male-typical call preceding the female-typical call. Cotton-top tamarins also have a predator alarm call; but after the threat is over, monkeys combine the alarm call with a relaxed contact call in a sequence that seems to act as an all-clear signal. The pygmy marmoset, the world's smallest monkey, shows several combined calls as well. Unmated monkeys often combine a long-distance territorial call with a food call when they find a highly preferred food, as though they are advertising this discovery of preferred food to potential mates. In contrast, already-mated animals give only the food calls. However, these examples lack the capacity to generate an infinite number of combinations, which is a hallmark of human language.

The best example of generative grammar in animals comes from a highly unlikely source, the black-capped chickadee, a small North American bird. The "chick-a-dee" call (after which the bird is named) is composed of four note types that can occur in one of two sequences (A) (D) or (B) (C) (D), where the note type in parentheses can be repeated any number of times (for instance, AADDD, BBCCCDDD, or BCDD), thus creating a generative grammar. Although the structural description of the chickadee grammar is simple and elegant, we have no evidence that each separate structure or even the two general forms (A) (D) and (B) (C) (D) have distinct meanings. This statement is perhaps unfair to chickadees, since Chomsky has argued that grammar is independent of functional reference (as in his famous sentence, "Colorless green ideas sleep furiously," which is well formed grammatically but creates semantic nonsense). More important, chickadee grammar lacks sequences of phrases (AADDDD, BBCCDDD) or embeddedness (AAA [BBCCCDDD] DDDD), which are aspects of human generative grammar.

Prevarication has until recently been difficult to document in nonhuman animals, but new reports on great apes provide extensive evidence of deliberate deception. (However, it seems to me a bit shaky to define language based on prevarication, for it makes our politicians and lawyers the most skilled language users!)

Songbirds born in one dialect group can learn the dialect of another population, and in some cases birds can even learn to sing the songs of different species if provided with an appropriate tutor. Mothers of cross-fostered rhesus macaques and Japanese macaques can learn to respond to the calls of their foster offspring, vervet monkeys learn to respond to the alarm calls of superb starlings, and ring-tailed lemurs can learn to understand the alarm calls given by Verraux's sifakas. All of these examples illustrate *learnability* and go well beyond the species-specific learnability that Hockett thought important.

Reflexiveness or metacommunication is difficult to establish in nonhuman animals, but a male chimpanzee who shakes a branch at a female to attract her attention to his erect penis and spread legs might be an example.

Hockett's final design feature is *duality of patterning,* the combining of sounds into words and words into phrases to have different meaning according to order (such as *god* and *dog,* or *boy, bit,* and *dog* to form "The boy bit the dog" and "The dog bit the boy"). Some New World primates (cotton-top tamarins and capuchin monkeys) combine individual elements or calls into sequences that retain the meaning of the individual elements much as a phrase or sentence does. These monkey examples always occur in a fixed sequence (such as ABC); thus we do not know if the order of elements (ABC) is semantically different from the sequence (CBA) or whether the monkeys would interpret both equivalently. Dolphins and great apes can respond appropriately to reversed word order such as "Take the ball to the chair" and "Take the chair to the ball," and both dolphins and bonobos can respond appropriately to sequences they have never experienced before, suggesting an understanding of duality of patterning.

Tetsuro Matsuzawa has trained a chimpanzee, Ai, to select the features that make up the lexigrams she uses for communication (Figure 8.2). Ai can use the same component (say, a small circle) in combination with other components to create different labels for different objects. Thus, nonhuman animals display a capacity for duality of patterning even if it has not been observed in natural communication.

No single nonhuman species seems to incorporate all of Hockett's criteria for language in its communication system. However, it should be obvious that at least one example of each of the design features can be found either in the natural communication of a species or as a potential that can be developed by specific training. If one accepts Hockett's design features as potential criteria for language, obvious continuities can be found between human and nonhuman animals.

Production and Perception of Speech

Humans are thought to be unique with respect to speech production and perception. As noted above, chimpanzees differ anatomically from humans, with only adult humans having structures for the full range of human speech. In adult humans (but not infants) the larynx is relatively low in the throat, creating the two-tube supralaryngeal resonator necessary to produce the vowels [I], [a], and [u] (as in *see*, *saw*, and *sue*). The much higher larynx in chimpanzees prevents them from producing these vowels, but they should still have the anatomic capacity to produce many other vowels. Chimpanzees also should be able to produce most of the consonant sounds used in English, though it may be difficult for them to produce [g] and [k].

Although some view the deficiencies of chimpanzees in vocal production as quite severe, I find remarkable the similarities between human be-

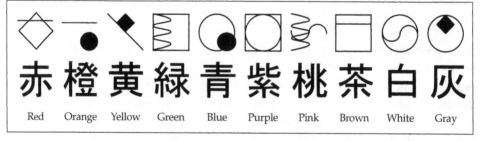

Figure 8.2 Some of the lexigrams used by Ai in the computerized communication system. Ai is able to construct the lexigrams from individual components; for example, choosing a square and a circle to form the symbol for *purple*, and a square with a horizontal line to form the symbol for *brown*. (From Matsuzawa, 1996. Reprinted with permission of Cambridge University Press.)

ings and chimpanzees in the potential to produce many speech sounds. Since some human languages have very few phonemes (the languages of Rotokas and Mura both have three vowels and eight consonants), in theory it should be possible for chimpanzees to create a spoken language recognizable to humans. It is surprising that chimpanzees do not speak, given their many anatomic parallels to humans. We cannot explain that nonperformance solely on the basis of the anatomy of the vocal tract. The failure of chimpanzees to speak must be due to differences in the motor control system, the motivational system, or the cognitive abilities of chimpanzees. Their elaborate visual-gestural communication system may be enough to meet their needs. If we look closely at the linguistic abilities of apes who sign or use computer symbols, there appears to be an upper limit on vocabulary of about 150–200 signs or lexigrams, and a limit on syntax of sequences of two or three items. These findings suggest fundamental cognitive differences between human and nonhuman apes that limit the development and expression of speech and language in the latter.

Humans were also thought to be unique in perceiving sounds as speech. It has been impossible to find acoustic invariants in speech that provide specific cues for sound classification, owing to the extensive coarticulation between a consonant and its preceding or following vowel. Thus, as we say the syllables *dee, daw,* and *dew,* there is no common acoustic feature in the physical structure of these sounds to indicate unambiguously the initial consonant [d]. For many years it was thought that only speakers of language could extract these invariant features based on the speakers' ability to produce these sounds.

A special type of perception, termed categorical perception, was hypothesized to be unique to language speakers. If presented with synthetic speech sounds that vary across a continuum of voice-onset time (from *bah* to *paw*), human subjects typically sort these sounds into two distinct categories with a sharp boundary between adjacent stimuli on the continuum. In traditional paradigms, humans could not discriminate different sounds from within a category (hearing both as *bah* or as *paw*), but they easily discriminated sounds from two different categories.

Many studies from the mid-1980s on have undermined the view that categorical perception is unique to speaking humans. First, preverbal infants discriminate speech sounds categorically. Then, several studies demonstrated that nonhuman species such as chinchillas and macaques categorize synthesized speech sounds as humans do. Japanese quail have been trained to learn a concept of [d] by being trained to identify the syllables *di, da*, and *du*, and then showing generalization to [d] associated with eight novel vowels. These studies show that the ability to discriminate human speech sounds is phylogenetically very ancient. Since it is unlikely that species such as quail, chinchillas, and macaques have ever been under significant selection pressure to discriminate human speech, the more likely explanation is that during human evolution our ancestors exploited longstanding, highly conservative perceptual capacities in constructing our spoken language.

Nonhuman species also can perceive their own sounds categorically. Mother house mice categorically discriminate between two different types of ultrasound produced by their pups, and swamp sparrows show categorical discrimination of song-note duration. Japanese macaques have seven different variants of *coo* vocalizations that they use in different contexts, and they discriminate between variants of these sounds categorically. In the cotton-top tamarin, our team found eight variants of chirp vocalizations used in different contexts, and the tamarins readily discriminated between the most similar of these variants. Nonhuman primates often have subtle variations in calls that might not be readily discriminated by human observers; these variants are associated with different contexts. Further, primates segment their calls into separate categories, providing further evidence of a discrete communication system. Careful acoustic analyses coupled with studies of the contexts of vocalizations and appropriate perceptual studies find that many nonhuman species have complex variations of sounds that they perceive in discrete categories, even if human scientists cannot always hear the differences.

The pygmy marmoset categorizes two forms of trill on the basis of duration. Short trills are used as friendly contact calls, and longer trills appear to be used in aggressive or fearful contexts. When tested with synthetic trill vocalizations varying continuously on call duration, pygmy marmosets showed distinct categorization. Later we found that pygmy marmosets also showed within-category discrimination if they were presented with synthetic calls representing familiar individuals. Thus, if a monkey heard one call synthesized to mimic Sally's trill and one synthesized to represent Mark's trill, the monkey readily discriminated between them. Whether one finds categorization or within-category discrimination depends on the demands of the experiment and the methods used.

Upon reflection, the idea of categorical perception as a distinctive criterion of language is odd. Essentially, listeners appear to ignore much of the acoustic information (age, sex, individual, intonation contours) in a signal when they sort speech sounds categorically. Why should listeners ignore this additional information? From an evolutionary perspective it makes sense for humans and nonhumans alike to attend not only to the linguistic or communicative content of a signal, but (since communication is a social transaction) also to nonlinguistic information: gender, individual identity, geographic origin, and emotional state. Whether or not this nonlinguistic information is useful depends on the context of communication. If I am in a dark theater and someone yells "Fire," I attend to the linguistic aspects and ignore other information that might be in the signal. But if in the same theater I hear the phrase, "Kiss me, I love you," I do not respond to the linguistic content without first identifying the sex, individual identity, and direction of the sound. Different social contexts require different types of information processing. Much research on human speech and language, as well as on animal communication, has attended to the formal structure of words, phrases, or calls and the semantics or meaning of these calls. Few researchers on the origins of language have focused on the social uses of communication, but it is precisely these social functions of speech and communication that play an important role in understanding the evolution and development of language.

Parallels in Nonverbal Communication

A significant portion of human communication is nonlinguistic. We use gestures, facial expressions, grunts, moans, and other nonverbal sounds along with language. If you ask me how I am and I reply "Fine" but have a slumped posture, a sad face, and a flat intonation to my voice, you will determine the reality of how I feel from my nonverbal signals, not from my language. In normal conversation we monitor not only speech, but nonverbal cues.

Many nonverbal cues are continuous between human and nonhuman species. Eugene Morton has built on Darwin's principles of emotional expression to develop a system of motivational-structural rules that appear consistent across many species. Aggressive animals give low-pitched noisy calls and make themselves look bigger by fluffing their feathers or fur. Submissive or fearful animals give high-pitched calls and sleek their feathers or fur to appear smaller. When humans talk to infants, we adopt a high-pitched voice and often stoop to make ourselves appear less threatening; we threaten each other with low-pitched growls.

Research on how humans communicate with dogs, horses, and other working animals finds consistency across many cultures and language types in the signals used. A series of short, upwardly rising calls arouses an animal to activity ("giddyap, giddyap"), a slowly decreasing pitched sound calms or slows an animal ("whoooa") and a sharp, staccato bark leads to a rapid end of activity ("stop!"). Human parents use similar sounds to control the state of their infants.

Jan van Hooff, comparing the facial expressions used in different contexts in monkeys, apes, and humans, has found remarkable similarities across species. Macaques and chimpanzees have a large number of different facial expressions used in communication. Chimpanzees can easily discriminate between pictures of other chimpanzees with different facial expressions, and macaques can discriminate between happy and sad human faces—although they seem to use different cues than humans do to make this discrimination. Thus there are clear continuities in facial expressions and nonverbal vocalizations between monkeys, apes, and humans.[2]

Is There a Language Instinct?

A fundamental argument for the uniqueness of language is its rapid development. The complexity of language combined with the early and rapid ability of human infants to acquire the fundamentals of language (words, syntax, and meaning) and the similarities of language development across cultures has suggested to many that language is an instinct—a response that is largely hereditary and unalterable (Webster's Dictionary). Species-specific modules are said to exist in the human brain designed to promote particular language abilities. While there is some acknowledgment of the value of learning and experience in language acquisition, these factors are often dismissed as minor. In this section I review some of the evidence for a "language instinct" and suggest alternative ways of viewing language development. To the extent that language development is a product of general learning and cognitive processes that can be found in other species, language itself will have greater continuity with other species.

The argument for a "language instinct" relies on minimizing the diversity of different languages and finding universal features in development across language groups. There are some near-universals in language. All languages use stop consonants, and most use three places of articulation as represented by *ba* (lips closed, then opened), *da* (tongue placed behind the teeth) and *ga* (tongue raised at the back of the mouth). Try pronouncing these sounds and notice the position of your tongue, lips, and teeth. All languages use voicing as a contrast: *ba* versus *pa*. In both the lips are closed to make the sound, but in *ba* the larynx is vibrated at the same time as air is pushed through the lips. With *pa* the larynx is vibrated after the air is pushed through the lips.

Regardless of the number of vowels in a language, most languages use the three vowels [I], [a], and [u] that chimpanzees cannot produce. At the same time, there is tremendous diversity in human speech sounds. The UCLA Phonological Segment Database shows a total of 558 consonants, 260 vowels, and 51 diphthongs (two or more vowels combined as in "our"). Obviously no one language uses all of these sounds (!Xu, a South African language, uses 141 sounds and the languages of Rotaka and Mura utilize only 11 sounds), but the fact that 869 different sounds have been identified across modern spoken languages suggests a need for great flexibility in the learning of speech.

The psychologists Elizabeth Bates and Virginia Marchman have argued that there are few grammatical universals in language development. While they accept the fact that children are predisposed to communicate with others and to learn language (as birds are predisposed to learn song), they find little evidence of a universal sequence of grammatical development that follows a fixed timescale across languages. Languages with grammatical forms different from English, such as Turkish, Russian, and Italian, give different importance to word order, the relative complexity of verb endings, and whether case (subject, object, indirect object) and gender are coded in nouns. Across these various languages, children seem to learn first the aspects of grammar most important to their native language rather than following a universal trajectory. Even within English speakers, Bates and Marchman argue that there is no all-inclusive pattern of development. The onset of production of words, phrases, and acquisition of negative or passive forms is highly variable across individuals and does not seem related to any major neurological markers of development.

Critical Periods in Language Development

Many linguists and psychologists have argued that language must be learned within a critical time period, with puberty serving as a cutoff. Sources of data are children with brain damage at different ages, and socially isolated children deprived of language contact until after puberty.

These examples are not ideal because a socially isolated or brain-damaged child may have suffered retardation or had neurological problems that hindered language learning. Social isolation may have prevented a child from acquiring the social skills that are necessary prerequisites for language learning.

An alternative method is to study bilingualism. The psychologists Judy Johnson and Elisa Newport designed a brilliant study looking at native Chinese or Korean speakers who moved to the United States at different ages. At the time of testing all had spoken English for at least five years and had resided in this country for at least three years. The scores on a test of English grammar appeared to support a critical-period hypothesis. Those who learned English before age eight years had scores equal to native English speakers. There was a progressive decrease in performance with increasing age of first exposure to English. However, even the worst performance was at a 75 percent level. This is quite different from the completely disrupted speech of isolated human infants or the disrupted performance of songbirds deprived of auditory input prior to their critical period. There was also great individual variation, with some people in the late-exposure condition performing as well as those exposed to English at the age of eight.

Other results countering a critical-period hypothesis come from an indigenous population of ten thousand people in the Northwest Amazon on the border between Brazil and Colombia. Owing to strongly localized languages and a high degree of marriage between people from different groups, each individual knows two, four, or more languages. Minimally, a child starts by learning the separate languages of the mother and father—and with people from several language groups living in the same village, the child may learn more. A young adult seeking a mate might be lucky enough to find a partner speaking a maternal or paternal language, but more often mates must learn each other's languages (and then those of their in-laws!). The 25 languages identified in this area come from four different language families; thus, many of the languages are distinct in grammar and sounds (it is not like learning one Romance language and generalizing to others). The study reported that people did not use a language until they felt they knew it well. While adults might speak a new language with an accent, children spoke a newly learned language without accent. It seems that language fluency is valued and that both adults and children can attain a high degree of fluency in newly learned languages (though children speak without an accent).

Here is a more personal alternative explanation of apparent critical-period effects: I first learned French in high school, then German at college, Spanish when I was in my late thirties, and Portuguese in my mid-forties. I am fluent enough to be understood when lecturing in these languages, and I can find my way around in the countries where each of these languages is spoken, but my speech is far from perfect. German, which I studied most intensively, is my most fluent language, especially in social conversation. I learned Spanish and Portuguese for research-related purposes in the midst of a busy professional life. At the time I was less tolerant of critical feedback than I was as a student, and I suspect that others were less inclined to correct me at my relatively advanced age. My diminished fluency in Spanish and Portuguese (compared with German) can be explained by my very different mode of learning these languages. I had less time to do rote drills, abbreviated practice, and less consistent feedback. What might appear to be a critical-period effect in my second-language learning might be explained just as well as an artifact of different cognitive styles applied to language learning at different ages, and by the different demands placed on learners. The amount and quality of learning coupled with social reinforcement may be more important than age in influencing second-language fluency.

Learning may play a more significant role in human-infant language learning than previously thought. Studies have shown that young infants are successful in identifying statistical regularities in speech input and may use these abilities to learn language. A characteristic of a word is the frequent co-occurrence of the sounds and syllables that make up the word. When eight-month-old infants were presented with a two-minute string of nonsense syllables, the infants learned to recognize triads of sounds that were associated 50 or 70 percent of the time. If eight-month-old infants can learn to recognize "words" by extracting statistical transition probabilities from two-minute sequences of nonsense syllables, how much more might they learn through normal exposure to speech accompanied by social reinforcement?

Parents or infant caregivers are principal contributors to language development. The amount of language input a child receives influences her or his rate of language development. Data suggest that abused infants actually score above normal age levels in language production, whereas socially neglected children lag behind, suggesting that intensity of social interaction may encourage language development.

A study of fraternal twin sisters in France found consistent differences in language development. One twin had a preferred attachment to her father, the other to her mother. Each twin showed a high rate of reciprocal interaction with the preferred parent. The father was often away from home and had less contact with both twins than the mother, but nonetheless influenced the language development of the twin who preferred him. The twin who interacted most with her mother produced words sooner. The twins were reared in identical gross environmental conditions with equal opportunity for interacting with parents, but the fact that language acquisition was linked to the choice of the preferred parent suggests the importance of the nature and quality of social interaction in language development.

These examples hint that language learning need not be constrained by critical periods and that environmental influences such as learning and social interaction are crucial components of language development. A proponent of a "language instinct" might argue that a child has a strong internal motivation to learn a language and that the ability to extract statistical probabilities from just two minutes of speech input requires some specialized mechanism. That may be, but there appears to be much more to language development than just the activation or maturation of innate brain mechanisms. Social and cognitive mechanisms have at least as important a role in language development as biological mechanisms.[3]

Communication Skills in Nonhuman Animals

One of the most glaring exceptions to the idea that nonhuman primates are highly intelligent and flexible animals appears to be the development of communication skills. Studies of birdsong show that young birds must be exposed to the songs of adult tutors during development, and that subsequent practice prior to a male bird's first breeding season is necessary for a bird to develop normal song. After this practice period, song becomes "crystallized" and further modification is unlikely. A similar argument has been made that children must be exposed to the sounds of language early in life and that infants "babble" to practice the phonemes and words of their language.

Many nonhuman primates produce calls that are the functional equivalent of birdsong: the loud calls that are given only by males of many Old World primates, the songs of both sexes of gibbons and titi monkeys, and the long calls of marmosets and tamarins all function to attract and retain mates, repel same-sex intruders, and defend territories. These calls have been assumed to be genetically fixed owing to evidence from studies of hybrid gibbons who have songs that are a mix of both parental species, studies that use the structure of long calls in taxonomy and thus imply conservative structure, and studies showing that deafening and isolation of young primates has little influence on loud call, long call, or song development.

This evidence from monkeys suggests that, with respect to vocal development, monkeys are different from songbirds and humans. However, a communication system really involves three distinct components: production of signals, using these signals in appropriate contexts, and responding to the signals of others. Recently, Seyfarth and Cheney reviewed developmental studies and found that the production of signals is relatively fixed across primates (human and nonhuman alike) with only some modification within constraints. They found greater flexibility in the ontogeny of the usage of signals and the most flexibility in how monkeys develop abilities to respond to vocal signals. They suggest that studies of nonhuman primates may show strong parallels to human language development when all aspects of communication are considered.

For several reasons it has been difficult to find plasticity in primate vocalizations. Many of the calls that have been studied are used in contexts where plasticity might be maladaptive. Calls used in territory defense or to rally a group against intruders are strongly species specific and conservative. Calls given by infants separated from their caregivers should not be plastic or flexible, because the very survival of infants depends on accurate perception of these calls. Alarm calls also show little evidence of plasticity. In life-or-death situations, highly stereotyped calls should be much more adaptive than highly variable calls.

Recently we have learned that the subtle structure of song and other bird vocalizations can be influenced by interaction with social companions. European starlings share the songs of close social companions, and when those companions change, starlings change their songs to match those of their new comrades. Budgerigars in groups share similar calls. Black-capped chickadees live on territories for part of the year, but form social flocks to survive the winters of the Northern Hemisphere. Chickadees converge in the structure of some aspects of their call notes within a week after forming flocks in late autumn. The American goldfinch uses a variety of call notes, and when a male and female mate, each gives up some of its own calls and acquires some of the calls of the new mate. Each of these species lives in year-round social associations with other animals, in contrast to the territorial and seasonal association found in the birds studied most extensively.

Since most monkeys live in year-round social groups, vocal plasticity may be most obvious among vocalizations used to communicate about social relationships, especially when those relationships change. Thus, it may be more profitable to study how calls are used to maintain affiliative relationships among social companions than to look at infant separation or predator alarm calls. If language evolved to serve social functions and develops within a social context, then studies on the development of affiliative social communication in nonhuman primates might be the most productive place to start.

Trill Vocalizations in Pygmy Marmosets

Our own research has focused on affiliative signals of marmosets and tamarins, and much has been done with captive animals living in natural social groups. We can more easily record the often quiet and subtle calls used in social interactions in captive settings than in the wild, but we do field studies as well to verify that what we see in captivity is realistic.

Pygmy marmosets produce high-pitched frequency-modulated calls throughout the day. Individual-specific features of the calls allow other group members to recognize familiar individuals. Monkeys take turns calling; frequently each group member calls before anyone else calls a second time, with a regular order of turn-taking. As individual pygmy marmosets move farther away from other group members, they shift the acoustic structure of these trills to improve sound localization. Trills are not simply reflexive. Pygmy marmosets use them in appropriate sequences and alter the structure to make them more cryptic or more easily localized. By using these calls continuously throughout the day, pygmy marmosets have a "social map" of their group, always knowing who is calling and where in the habitat the caller is located.

We have studied the ontogeny of trill vocalizations from birth through the second year of life in monkeys housed under seminatural captive conditions. Call structure changed with age, so calls were not completely genetically determined. Although some of the monkeys changed call structure in ways predicted by a model of physical maturation, other monkeys did not, suggesting that vocal changes could not be related to simple physical growth. We found little evidence that monkeys were developing the increased stereotypy of vocal structure that a critical-period model requires. It appeared that our monkeys, like goldfinches, budgerigars, and chickadees, might alter vocal structure throughout their lives.

We predicted that changes in social structure would induce changes in trill structure. In one study we recorded trills for several weeks from four males and four females while they were living in family groups, and then we formed four breeding pairs from these animals with the male and female in each group being unrelated and unfamiliar. We continued to record trill vocalizations for six weeks after pairing, and recorded six of the monkeys for three years after they had been paired. We measured several acoustic variables in the trill vocalizations of each individual.

Prior to pairing, there was little change in the structure of trills over a two-month recording period. But after pairing, three of the four newly mated pairs changed the acoustic structure of their trills to produce trills more homogeneous between mates than before pairing. In the fourth pair, the trills were nearly identical at the start and no change was seen with pairing. Monkeys that differed the most in trill structure prior to pairing converged the most after pairing. When we recorded monkeys three years after pairing, there was still clear convergence between mates even though some aspects of trill structure had changed. Thus, when pygmy marmosets acquired a new mate, they quickly modified the structure of their trills to converge with those of their new mate, and they maintained convergent trill structure for up to three years.

Similar results have been found in goldfinch calls, in European starling song, in the foot-drumming signature of kangaroo rats, in greater spear-nosed bats, in bottle-nosed dolphins, and in dwarf mouse lemurs. Parallel studies of humans show that we modify our language when joining a new social group. This "optimal convergence" is found in humans of all ages from young children through adults. We speak differently to children than to adults, and many social groups mark themselves by special words or phrases, a form of social "badge." (The jargon used by scientists of every discipline is a type of social badge.) When we join a group, we indicate our interest in joining by altering how we speak; but although we modify our speech to indicate solidarity with a group, we still maintain our individual distinctiveness as well. Each of these examples from birds through bats to dolphins, pygmy marmosets, and humans represents a case of "optimal convergence" and suggests that this type of vocal plasticity has a long evolutionary history.

Babbling in Pygmy Marmosets

Babbling is thought to help human infants develop language skills. It has been thought to be uniquely human, since no evidence of babbling was previously observed in monkeys or apes. The closest equivalent in nonhuman species has been witnessed in some songbirds. Near the start of the breeding season, year-old birds begin to produce many highly imperfect versions of song. Rapidly the songs become increasingly more organized and more closely approximate adult song, until crystallized song appears at just about the time that males compete for territories and females. Although this song practice has often been compared with babbling in human infants, there are major differences. In many songbirds males—but not females—sing, and the "babbling" period corresponds with puberty. Furthermore, song is only one of a large number of vocalizations in a bird's repertoire.

In human children, babbling begins early and is completed long before puberty, it is not specific to males, and it includes a large number of phonemes used in speech. Infant pygmy marmosets are extremely vocal, producing long sequences of calls lasting several minutes. During the first twenty weeks of life, pygmy marmosets show a number of parallels to babbling in human infants.

1. Babbling is universal in both species.
2. Babbling begins early in life. Human infants start babbling by seven months of age and pygmy marmosets babble as early as the first week of life, with all infants babbling by the fourth week (Figure 8.3).
3. Babbling is rhythmic and repetitive. In both species the same sounds are repeated several times in a row before an infant switches to a new sound.
4. Babbling uses a subset of adult sounds. Typically, human infants use only about 50–70 percent of the sounds of their parents' language, and pygmy marmosets use 56 percent of the call types that adults produce.
5. Human infants produce sounds that are recognizable as phonemes from speech. Of the 21,000 calls that we analyzed from pygmy marmosets, 71 percent were adultlike and another 19 percent were recognizable as variants of adult vocalizations.
6. Babbling has no obvious referent. Human infants produce sounds

Figure 8.3 Male pygmy marmoset carrying twin infants. Infant marmosets babble frequently at this age and are more likely to be in social contact with other group members while babbling. (Photo by Carla Y. Boe.)

with no evident relation to objects or events in the environment. Pygmy marmoset babbling consists of many different call types produced in close juxtaposition: an alarm call might be followed by a threat vocalization, then by an affiliative call (Figure 8.4).

7. Babbling is a social act. Caregivers respond to infants when they babble, providing reinforcement for still more babbling. Pygmy marmoset caregivers are highly responsive to babbling infants, with much higher rates of approach to the infant, of grooming or contact behavior, and of carrying when infants were babbling than when infants were not babbling.

Thus, each of the seven major features of babbling in human infants has parallels in the vocalizations of infant pygmy marmosets. Furthermore, babbling seems to improve the quality of infant marmoset calls. Infants who were the most vocal early in development had the most adultlike trill structure at five months of age.

Babbling is not unique to captive pygmy marmosets, but has also been

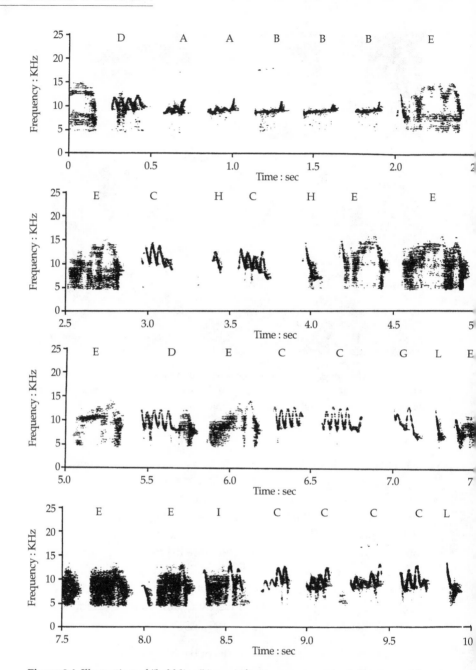

Figure 8.4 Illustration of "babbling" in an infant pygmy marmoset. Frequency of sound is on the vertical axis, and time on the horizontal axis. The figure shows a 10-second sample from a much longer babbling bout. The letters above each note indicate the type of call being produced. Note the frequent repetition of notes and the rapid transition from one note type to another. (From Elowson, Snowdon, and Lazaro-Perea, 1998a.)

observed in wild populations of pygmy marmosets and common marmosets. Interestingly, we found nothing similar to babbling in tamarins, nor has anything comparable been reported in other monkeys or apes.

Ontogeny of Cotton-top Tamarin Food Calls

Both marmosets and tamarins are noteworthy because fathers and elder brothers are actively involved in carrying young infants. Males play another role as well, in that they share food with infants during the weaning period. This activity provides a possible mechanism for adults to teach infants about appropriate food and about the vocalizations associated with food. Cotton-top tamarins have two chirplike calls used when feeding. During food sharing the male who offers food to the infant vocalizes, at the same time producing a rapid sequence of the chirplike calls adults use when feeding (Figure 8.5). Infants are able to receive food only if and when the adult calls. Frequently during food sharing, other group members orient toward the male with the food and give food calls as well. Infants with whom a large amount of food is shared eat independently and use food vocalizations during feeding at an earlier age than other infants. Thus, food sharing creates a context for infants to receive solid food, to recognize what is edible, and to learn the calls associated with feeding. Food sharing is a form of social information donation that facilitates infants' transition to eating solid food and to learning the appropriate use of calls associated with food.

After infancy, tamarins produce imperfect forms of food calls and throughout the entire time that they live in family groups, they continue to produce these imperfect calls without noticeable improvement. However, immediately after tamarins are removed from family groups and paired, they begin producing adult forms of calls. The greater the amount of affiliative behavior between newly formed pairs, the faster the transition to adult food calling. Since postpubertal adults in tamarin families are reproductively and socially subordinate, it is plausible that the continued use of infantile forms of calls long past puberty signals subordinate status (Figure 8.6). The rapid change in vocal production with change in social status supports this hypothesis.

Thus, social companions can both facilitate and inhibit development and change in vocal structure and usage in marmosets and tamarins, but what are the mechanisms of social influence? Most of us dread making tele-

Figure 8.5 Cotton-top tamarins sharing food. Fathers and elder brothers do most of the food sharing with infants and in the process give rapid strings of food calls. (Photo by Carla Y. Boe.)

phone calls when traveling in a country where we are using a second language. It is much easier to understand someone if we have simultaneous visual and vocal cues. Learning about communication might occur more rapidly were cues to appear in more than one modality. Similarly, engaging multiple sensory modalities might enhance language acquisition. Further, social interactions provide a focus of attention. We are more likely to attend to the behavior of a social companion than to an inanimate object, and a social companion can direct the attention of a learner to the relationships between the sound and the object, as the adult male tamarin does when vocalizing as he offers food to the infant. Finally, social companions provide reinforcement, either direct (as in the case of the male tamarin sharing food with an infant) or indirect (by providing praise or physical contact). The example of cotton-top tamarin food sharing illustrates how all three mechanisms can operate simultaneously.

Figure 8.6 Cotton-top tamarin vocalizing. Young cotton-top tamarins may learn appropriate use of calls through interactions with adult group members; however, some aspects of call structure and usage appear in infantlike ways in nonreproductive adult tamarins, perhaps to communicate subordinate status. (Photo by Carla Y. Boe.)

Strong developmental parallels exist between language development in humans and the ontogeny of vocal communication in nonhuman primates if we focus on appropriate comparisons. There are constraints on the types of sounds that each species can produce, but there is also room for plasticity, especially in adjusting to the communication of new social companions. Acoustic structure can be modified through social interactions, and organisms have some capacity to produce new variants of species-typical vocalizations at all ages. Social interactions can reinforce or inhibit developmental processes. Babbling in marmosets and humans provides a means of practicing adult vocalizations. In marmosets the amount of practice directly correlates with the rate of developing adult vocal skills. Although the outcomes of human language and marmoset and tamarin calls are quite different, the developmental processes share many common features.[4]

The Evolution of Language from Animal Communication

How could we move from communication systems in nonhuman primates to human language in a manner consistent with evolutionary principles? Arguments that humans are fundamentally different from nonhuman animals either set the stage for creationist explanations or simply avoid the attempt to develop a persuasive evolutionary argument. Bickerton's proposal of a single-gene mutation is, I think, too simplistic. Too many factors are involved in language learning—production, perception, comprehension, syntax, usage, symbols, cognition—for language to be the result of a single mutation event.

Many complex phenomena can be explained parsimoniously through bottom-up principles that result from self-reinforcing social interactions among individuals. The complex organization of a colony of social insects can be modeled from the sequences of interactions among sets of two individuals within the colony. Similar self-organizing models have been developed for explaining dominance hierarchies and reciprocal altruism in nonhuman primates. It seems possible that complex communication and language may be constructed in the same way, through the selective reinforcement of simple interactions between individuals. For example, although most birds need song tutors to acquire normal song, a group of four young zebra finches raised together will develop normal song without a tutor. Deaf children develop complex sequences of gestures to communicate with parents in the absence of specific teaching, and an isolated population of deaf people in Nicaragua has created its own sign language system.

Consider the available evidence for what nonhuman animals can do. Most of Hockett's design features appear in at least some nonhuman species, and perceptual capacities for speech sounds appear to be phylogenetically quite ancient. There are many parallels in ontogeny of production, perception, and usage between human and nonhuman species, with the caretakers of infants playing an important role in shaping communication skills. Great apes appear to express in their behavior some of the cognitive abilities thought to be necessary for language. Matsuzawa described the tool-use patterns of wild chimpanzees in a hierarchical analysis similar to those describing syntactic relationships in language. Archaeologists have argued that patterns of tool-making in early hominids illustrate a syntactic ability comparable to language. Thus we might find some evidence of the origins of syntactic ability in apes and hominids.

Studies of apes and dolphins trained to use or respond to arbitrary symbols suggest that specific learning principles can be used to train symbolic skills. With or without explicit teaching, infant primates rapidly learn a variety of social, food-processing, and predator-avoidance skills through observation of adults. We have shown how food sharing in cotton-top tamarins provides a context for learning about food and communicating about food, and how babbling in pygmy marmosets leads to the development of adultlike vocalizations. These mechanisms that lead to competence in monkey communication could lead to adult modes of production, usage, and comprehension in language as well.

But some data create problems for this scenario: there is little evidence as yet, in great apes, of the complex vocal communication with referential signaling found in other nonhuman primates. Yet apes (and some monkeys) have very expressive faces and can use gestures such as pointing to request objects. Gestural and visual signals may be more important for apes than vocal signals. Researchers of sign languages in humans have argued that sign offers a much more rapid rate of information transfer, since multiple signals can be provided simultaneously: hand shape, hand movement, speed and direction of movement, coupled with facial expression and relation of hand and finger movement to other body movements. Greater complexity and efficiency of communication may have first developed in the gestural-visual modality. Corballis in the 1990s revived a gestural-origins argument.

As great apes and humans also have highly overlapping binocular fields of vision, we are unable to detect visual or gestural signals that are more than 70–80 degrees to either side of the center of our faces. Furthermore, we cannot perceive visual signals or gestures in the dark, or in forests or thick grasses. Here auditory-vocal communication offers its advantages: we can communicate with companions who are out of sight while still directing visual attention toward a predator, a prey, or a valued social partner. By vocalizing we free our limbs for locomotion, or for carrying weapons or other objects. We can extend our activities into the night and into habitats that might not be usable with gestural communication alone. While the majority of terrestrial Old World primates and great apes may use gestural-visual communication more than auditory-vocal systems, those of our human ancestors who could exploit preexisting features of the auditory system to create a vocal language would have moved an important step beyond other primates. A rapid expansion of vocal communication would have led to more complex social and cognitive interactions with a greater number of individuals, which in turn would have led to increased cranial size and a more complex neocortex specialized for vocal production and auditory analyses.

Words could have developed from two sources. We have seen that some monkeys have different calls for different predators, and an adaptive response to the predator differs as a function of predator type. Precise warnings about predators have obvious adaptive value. Requests for objects constitute another route toward words. Young tamarins vocalize extensively in food-sharing settings, and capuchin monkeys and apes can request objects by extending an arm or by pointing. Human infants use grunts to indicate objects, and these grunts are subsequently replaced by words.

What is the adaptive significance of increased communication ability? Language evolution was driven by social forces. Those ancestors who could communicate most effectively about the location of food and shelter or about predators and how to avoid them would have had more reproductive success because of obtaining more matings and by leaving behind more surviving offspring. Dunbar (see Chapter 7) has argued that "gossip" is an effective way to maintain social relationships among the increasing number of individuals found in human, as opposed to nonhuman, primate social groups. To the extent that increased social support provides increased survival advantages through finding food, avoiding predators, and managing relationships within a group, a more complex language is adaptive. Stanford (see Chapter 4) finds parallels between cooperative hunting in chimpanzees and humans. Although hunting does not provide a major component of the diet, successful hunting appears to be important—a critical criterion of mate choice. Thus, if increased communication abilities led to better coordination of cooperative hunting, and hunting success led to increased reproductive success, this would be a selective advantage to developing language. Language skills that allowed human ancestors to hunt more effectively, to exploit additional habitats, to be active after dark, and to manage social relationships in increasingly larger groups would have conferred a distinct advantage to early hominids over extant apes.[5]

The developmental and adult data from nonhuman animals support a bottom-up empiricist approach to language origins that establishes language as the current end point of a continuum of communication abilities. While humans do have species-typical adaptations for speech and language, it is not necessary to hypothesize special perceptual abilities, special cognitive abilities, or special brain structures to support language. It is more parsimonious to view these as resulting from the increased survival value accruing from the complex communication abilities that have emerged during the process of primate evolution.

9

The Nature of Culture: Prospects and Pitfalls of Cultural Primatology

William C. McGrew

If other primates transfer skills and habits from one generation to the next, are we not justified in speaking of culture as well? Wild chimpanzee communities, for example, vary greatly in tool use and special forms of communication. This intergroup diversity suggests that we are not the only cultural beings.

ALL LIVING HUMANS exist in societies permeated by culture, and such cultural processes transcend our basic subsistence in the world of nature. Many North Americans still dress baby girls in pink and baby boys in blue not because of genetic or environmental constraints, but because it is the custom to do so.

But what of our ancestors? How far back in time can we trace this pervasive influence on our actions? It is easy to posit culture to past *Homo sapiens*, who were anatomically indistinguishable from ourselves even more than 100,000 years ago. The vivid images inscribed on the cave walls of Lascaux, Cosquer, and Chauvet in France and Altamira in Spain testify to the talents of that era. Similarly, our "cousins" the Neanderthals, who survived until as recently as 30,000 years ago, produced exquisite flaked stone blades (the Mousterian tradition) that suggest cultural norms.

When we seek to look further back in time, we become largely dependent on material culture in the form of stone technology; all else has perished and so is "invisible" to archaeology. Objects made of wood, horn, bark, fiber, leather, and even in many cases bone decomposed long ago. Luckily, the stone tools of *Homo erectus* left a distinctive record (the Acheulean tradition) of bifaced hand axes dating to about 1.8 million years ago. Even as long ago as 2.6 million years, earlier hominids made and used the simplest of flaked stone tools: the scrapers and choppers of the Oldowan tradition. Beyond that time, there is nothing that we can recognize as cultural—material or otherwise.

Yet it seems likely that all of these ancestral forms, and probably also their predecessors, the earliest australopithecines, used nonlithic materials and practiced cultural ways. How, in the absence of direct evidence, can we infer this? The best way available to us is to scrutinize the other living organisms with whom we last shared common ancestry. These are our closest relations—the primates, especially the great apes—with whom we parted evolutionary company only 5 million to 6 million years ago.

Primatocentrism?

Before we focus on primates, however, consider the following thumbnail ethnographic descriptions:

Case 1. Two communities live along the northwest Pacific coast of North America. One subsists largely on marine mammals, such as seals and sea lions; the members hunt in small, silent parties, roving widely. The other community focuses on fish, especially schools of salmon; its members hunt in big noisy groups and stay close to home. Both societies speak the same language, but with distinct dialects that differ even from clan to clan.

Case 2. Two populations live 250 kilometers apart, separated by high mountains. One group erects towers of glued sticks on a painted black mossy base, decorated in stereotyped style with black, brown, and gray snail shells, acorns, sticks, stones, and leaves. The other population erects woven-stick huts on an unpainted green mossy base, decorated with much individual variation, using fruits, flowers, fungus, and butterfly wings, of every color imaginable except a few shades of brown, gray, and white.

Case 3. Different groups colonized different types of forest, where they found little competition. The empty niches allowed remarkable innovation: these are the only societies known to build arboreal residences. Each group invented a range of efficient techniques to harvest staple foods, focused on the seeds of conifers. The processing techniques require social transmission from one generation to the next; youngsters deprived of such tradition would starve.

None of these case studies is of humans. The first is not a society of seagoing canoe-hunters of marine vertebrates, such as the Kwakiutl, but are orcas, or killer whales. The second is not a highland New Guinean horticultural society such as the Eipo, but a population of bowerbirds. The third is not a seafaring, exploratory colonizer of uninhabited islands, such as the ancestral Polynesians, but black rats.

The cautionary lesson intended here is that just because humans are primates, cultural processes need not be limited to primates, nor even to mammals. The order Primates is merely one of 20 in the class Mammalia, and other large-brained species, especially of whales, dolphins, and elephants, are likely candidates as culture bearers. Primates are the focus here because of their phylogenetic and ecological similarities to humans: we are land-dwelling organisms with skillful manipulative hands.

One way to distinguish human culture from that of other creatures, be they apes or songbirds, is to classify what they do as incipient or incomplete. Thus, what other species do may be called "culture," protoculture, quasi-culture, preculture, subculture, and the like. This labeling may create a somehow satisfying distinction between us and them, but it does not lend itself to scientific study. What is needed is explicit criteria that can be applied to the acts of other species, in order to test for cultural validity. To put it another way, saying that true culture is *by definition* human just passes the buck to the problem of defining humanness.[1]

Defining Culture

Edward Tylor, writing in 1871, is usually credited with the seminal definition of culture as "that complex whole which includes knowledge, belief, art, law, morals, custom, and any other capabilities and habits acquired by man as a member of society." Putting aside its restriction to humans only, at first glance it seems to ring true; but a second glance it is frustrating as an all-inclusive checklist. Suppose another species demonstrated everything on the list but belief—would we deny it cultural status? And how would we even recognize belief in another species? What exactly is a habit, which can range from absentmindedly pulling at one's earlobe to regularly attending mass? Even more problematic are the catchy aphorisms that sometimes appear in introductory textbooks: Culture is what humans do. Culture is the human ecological niche. Culture is what makes us human.

Another way to tackle these issues is to set out a list of features or characteristics of culture that can be confirmed empirically from data on the actions of a population of organisms. Table 9.1 sets out eight such criteria, as developed from the pioneering efforts 70 years ago of the cultural anthropologist Alfred Kroeber. Ideally, one would see the origins of a custom in its invention or novel modification, followed by its transmission from the innovator to others. Such patterning would become standardized, or even stylized, across the community's members and would persist after the innovator's departure or demise. Further, it would be handed down across generations and would spread across social units. Of course, such a cultural pattern should not be dependent on human influence (such as domestication), nor should it be determined by the constraints of the biotic or physical environment (such as lack of penguins in the Arctic). If a group of primates showed a behavioral pattern that met most or all of these criteria, it would seem churlish to deny them cultural status.

It is hard to expect all cultural primatologists to adhere to one set of imposed properties, so an alternative way to proceed is to seek general principles on which all can agree. By consensus, culture is learned rather than inherited. That is, organisms acquire culture through experiences that modify their acts and minds. Moreover, cultural learning comes from other, usually familiar members of the same species, rather than from other species or from natural forces. That is, culture is social, not individual. Further, culture is ongoing rather than time bound. Whether as a short-lived fad or as a long-lived tradition, culture repeats itself. Finally, culture is collective rather than solitary. That is, culture characterizes a group, whether as a clan or community or nation-state, so that its members are identifiable to one another, in contrast to other groups. This diversity provides the potential for cross-cultural comparisons that yield illuminating similarities or differences, or occasionally universals, such as incest avoidance.[2]

Perspective

Until recently, culture was considered the province of anthropology, especially of social or cultural anthropologists or ethnologists working in the present and of archaeologists working in the past. Now other disciplines

Table 9.1 Eight criteria for the presence of culture (adapted from Alfred L. Kroeber)

1. Innovation	New behavioral pattern is invented or modified
2. Dissemination	Pattern is transmitted from innovator to others
3. Standardization	Form of pattern is consistent, even stylized
4. Durability	Pattern persists beyond demonstrator's presence
5. Diffusion	Pattern spreads across groups
6. Tradition	Pattern endures across generations
7. Species valid	Not an artifact of direct human influence
8. Transcendent	Not determined by biophysical environment

take an increasing interest, especially in the possibility of nonhuman culture. This broadening of perspective is useful, but it also may sow confusion, because different disciplines usually ask different questions.

In the context of the current debate, most anthropologists ask *what* questions. What is culture? Is what apes (or dolphins or australopithecines) do culture? What constitutes a culture bearer? What elements (if any) are unique to human culture? What recoverable artifacts signify culture? The focus is on the phenomenon. Ethnographers record the culture, material and otherwise, of a given society. Ethnologists seek to explain the range of variation that exists across societies by diffusion, convergence, or differentiation. Archaeologists seek to recover and to interpret prehistoric evidence of the emergence and development of culture.

In contrast, most psychologists interested in the question of nonhuman culture ask *how* questions. How do individuals acquire and transmit culture? How (if at all) do species come to differ in these processes? How do patterns of cultural transmission reflect social and cognitive capacities? How does early experience in life affect the performance of cultural acts? The focus is on underlying mechanisms. Developmental psychologists track the ontogeny of cultural acquisition from birth onward, as in infant socialization. Comparative psychologists contrast cultural processes across species: monkeys versus apes or gorillas versus orangutans. The task or knowledge or mindset is less important than the means of transmission.

Finally, biologists tend to ask *why* questions? Why did evolution reward organisms that learned from one another? Why encode information in memes rather than in genes? Why share or withhold cultural knowledge from another? Why invest time and energy and take risks in being cultural? The focus is on function, that is, on survival value and reproductive success. The presence or extent of cultural capacity is amenable to cost-benefit analysis, as is any other trait. At some point, shadowing a proficient companion is more adaptive than persisting in individual trial and error. Similarly, the extra brainpower needed for social cognition does not come without costs, such as the higher levels of competition for resources that come from living in groups. Ethologists and behavioral ecologists worry less about the definition of culture or its component processes, and more about how it pays off in survival and fitness.

Clearly, each of the three disciplines—anthropology, psychology, and biology—needs to contribute to an integration if we are to take seriously something called cultural primatology.

236 · **William C. McGrew**

Methods

Confounding the issues of disciplinary differences are the techniques used. In studying primates, both human and nonhuman, anthropologists traditionally did observational field studies, psychologists did experimental laboratory studies, and biologists did either or both. Nowadays these distinctions have blurred on all fronts. Consider these current projects on chimpanzees: Michael Tomasello does controlled experiments on apes in the lab, and Richard Wrangham observes apes in nature. But also, Tetsuro Matsuzawa does carefully delimited experiments with wild apes, just as Frans de Waal observes social relationships in naturalistic groups in captivity. All four combinations of technique (experiment versus observation) and context (captive versus wild) are needed to gain the full picture.

The factor of setting is not, however, a simple dichotomy of captive versus wild, but rather a continuum, at least for cultural primatology. In the controlled setting of the laboratory, primates may be reared by mother, human surrogate, or no one; in human homes, infant apes may be brought up as if they were human. In both cases the effects of enculturation (or humanizing) can be tested. In typical zoos, primates live socially, well fed and protected from predators, but are deprived of the basic natural opportunities to range, forage, migrate, and the like. In wildlife parks, social life may be richer and the habitat may be more spacious and naturalistic, but ultimately the same constraints apply. Captive primates may be returned onto islands or into protected areas, where ideally they become more feral with each successive generation. Finally, primates may be studied in nature, either with provisioning (the artificial offering of prized food to tame them for close-range observation), or with varying degrees of human influence, depending on whether they raid crops, are hunted, or coexist in urban settings.

Comparisons of the same species of monkey or ape or prosimian across this range of settings may be instructive. For example, most species-typical vocalizations emerge everywhere, regardless of environment, whereas ability to copulate may be hopelessly compromised unless the social setting offers the right age and sex composition of companions. Some types of material culture, such as tool use, remain absent in some species (lemurs, for instance) no matter what the setting, or appear reliably in some species (such as nest building in great apes) whenever raw materials and models are available. Some patterns, such as social grooming, are ubiquitous in general but vary specifically in the smallest detail across social groups of the same species, such as matrilineal kin in Japanese macaques using particular techniques to remove parasitic louse eggs from one another's hairs, as reported by Ichirou Tanaka of the University of Tokyo. Finally, some differences across populations seem to have everything to do with custom and nothing to do with environment: unlike their counterparts in West Africa who crack nuts with stone or wooden hammers and anvils (Figure 9.1), chimpanzees in central and eastern Africa do not, even when the same species of nuts and the same raw materials are available to them. Teasing out the relative influences of the social, biotic, and physical environments on the daily lives of nonhuman primates is just as challenging to cultural primatologists as it is to anthropologists studying human primates.

Before tackling the ethnography of primates, we need to emphasize two key points about the quality of data. One concerns validity, often in the form of confounded variables. Capuchin monkeys in captivity are accomplished makers and users of tools, but capuchin monkeys in the wild are not. This contradiction is puzzling until one realizes that most captive studies are of the brown capuchin, while most field studies are of other

Figure 9.1 At Bassa Island, one adult chimpanzee uses a stone hammer and anvil to crack oil-palm nuts, while another watches closely. (Photo by A. C. Hannah.)

species of the same genus. Which, then, is the crucial variable, species or setting? Only further research can tell. Such a mismatch is not unusual: There are no captive mountain gorillas, yet most of what we know of the intimate behavior of the whole species in nature comes from these few "gorillas in the mist." Finally, many studies of social learning in great apes scrutinize their performance after exposure to human models, to see how much they can acquire in comparison with human learners. No one has yet studied the converse arrangement: how well humans can learn a task from ape models. Cross-species studies of acculturation remain inconclusive.

The second issue of data quality concerns reliability. Comparisons are only useful if the data points compared are of equal status. Most ethnographic data for primates fall into four general types: *anecdotes* are rare or even unique events, such as a wild chimpanzee's being seen (once!) to wear a necklace of knotted monkey skin; *idiosyncrasies* occur when an individual repeatedly shows a unique act or style, as did a captive bonobo who invented a new technique to fracture stone in order to make cutting tools; *habits* exist when an act is shown by several individuals, more than once, such as male capuchin monkeys who coordinate their hunting of squirrels; and *customs* should be reserved for patterns that are regularly shown by all appropriate members of a group (such as dominant males, parous females, weaning infants). Ethnographers of primates must record data of all four types, but it is invidious to draw contrasts between lineages, or groups, or populations, or species, when the data points have different levels of reliability.

Primate Ethnography

The idea that nonhuman primates might be cultural emerged in principle at the beginning of the twentieth century, when anthropologists such as Kroeber read the ingenious work of Wolfgang Köhler, the German psychologist who studied problem-solving in captive chimpanzees. However, decades passed before systematic fieldwork on primates provided findings that could truly be interpreted as evidence of culture.

The first such evidence emerged in the early 1950s in Japan, when students inspired by the ecologist Kinji Imanishi chronicled the behavior of Japanese monkeys. Especially important were the beach monkeys of Koshima Island and the snow monkeys of Shiga Heights. The invention of sweet-potato washing and wheat sluicing at Koshima and hot-spring bathing at Shiga Heights are well known, both in introductory textbooks and in popular imagery. The Japanese primatologists set high standards of detailed recording of the acts of individuals identified not just by age and sex, but by matrilineal kinship. All of the early studies involved provisioning, so that human influence was a confounding factor; for example, the monkeys were induced to enter the pools by apples thrown there by human observers.

More recent studies of free-ranging Japanese monkeys are not compromised by these factors: Koshima's monkeys have added to their diets novel food items that have nothing to do with humans, such as octopus. The most compelling new data come from Yakushima Island, where the monkeys neither have been provisioned nor raid crops, yet they show a distinctive diet—they are the only Japanese monkeys known to prey on frogs and lizards.

Beginning in the early 1960s, researchers fanned out from Europe and North America to Asia, Africa, and Central and South America to do field primatology. Few if any went seeking evidence of culture, though many were inspired by the ideas of Louis Leakey and Sherwood Washburn to seek models of human evolution. As multiple studies of the same species of primate, but at different sites, revealed variation in behavior, explanations for this diversity were needed. Most of the time—but not always—environmental variables were enough: rain forests offer different resources and challenges than do savannas.

One of the first taxa of Old World monkeys to be studied was baboons, in both savanna and desert. Groundbreaking research done in South Africa by K. R. L. Hall, in Kenya by Irven DeVore and by Stuart and Jeanne Altmann, and in Ethiopia by Hans Kummer, was followed by the work of scores of their students and colleagues. Kummer's desert baboons in arid Ethiopia showed different sociosexual behavior than did savanna baboons elsewhere. The former live in harem groups headed by a dominant male; the latter mate more promiscuously, with high-ranking males seeking exclusive but short-lived consortships with receptive females. (Still later research by Barbara Smuts showed that the groundwork for these partner preferences was laid in special friendships between males and females that involved preferential grooming and protection.) Kummer and his students from Zurich were able to tackle the nature and nurture of the differences between the desert and savanna baboons by concentrating on a hybrid zone along the Awash River, where the ranges of the two species overlapped naturally. Using field experimentation, they were able to augment observational data by trapping and then introducing members of one species into troops of the other species. Both inheritance and experience contributed to the outcomes, as for example desert males "shaped" savanna females to be harem members.

The best comparative body of research on material culture in captivity comes from brown capuchin monkeys studied in colonies around the world. All are adept at elementary technology, and their performances seem to match and outdo anything shown by great apes. In Gregory Westergaard's studies these New World monkeys, who are only distant relatives of the ape and human line, make and use tools of wood, bone, bamboo, stone, even metal. They use such tools as containers, thrown missiles, digging sticks, and to crack open bones—all tasks analogous to the evolutionary origins of human technology.

All of these problems to be solved were presented by human experimenters in the facilitating surroundings of the laboratory, and none of these accomplishments has yet been seen in nature. Furthermore, researchers of capuchins entertain lively debate as to the extent of intelligence that underlies these impressive performances. Some argue that the behavior is mindless and unintentional, the product of trial and error, as indicated by elementary mistakes the monkeys make again and again when they fail at a task. Others point out that the performances, as measured by success rates and speed of learning, are no different in many cases from those of larger-brained apes. Until long-term behavioral studies at close range of wild capuchins are done at several sites, it will be hard to make sense of the intellectual and cultural capacities of these primates. Luckily, such research is under way at locations such as Santa Rosa, in Costa Rica.

Each of the four species of great ape—bonobo, orangutan, gorilla, and chimpanzee—represents a different aspect of the prospects and pitfalls of cultural primatology. Each is like us in many sometimes uncanny ways, but only chimpanzees are now in a position to supply ethnographic data useful for systematic comparisons.

Bonobos are the least known of the great apes, being found in only one country, the Democratic Republic of Congo, and subject to only two long-term field studies, at Lomako and Wamba. Both are lowland rain-forest sites, but while Lomako's bonobos have never been provisioned and remain mostly unhabituated to close-up study, Wamba's have received consistent gifts for many years. Thus, when Wamba's bonobos more often go bipedal to carry away armloads of sugarcane, what are we to make of this contrast with Lomako? Is their upright locomotion an artifact of provisioning? Research on savanna-dwelling bonobos from the south of

Congo suggests much greater plasticity in the species than was suspected, but we have a long way to go. Only in the 1990s did bonobo fieldwork reach the level of chimpanzee fieldwork in the 1960s.

Orangutans, the only surviving Asian great apes, have been studied at a variety of sites on the large islands of Borneo and Sumatra. Their solitary, low-density, secretive life style makes them difficult to study, particularly because they are the most arboreal of great apes. The two best-known studies, at Tanjung Puting in Borneo and at Ketambe in Sumatra, are complicated by simultaneous, interacting research on wild orangutans and on released orangutans being rehabilitated in the same forests. Anne Russon's work on the rehabilitants at Tanjung Puting shows them to be clever idiosyncratic imitators of human actions, to the extent of stringing hammocks and tending fires, but such ingenuity does not meet the consensus conception of ape culture outlined above. On the other hand, the fieldwork of Carel van Schaik and colleagues at Suaq Balimbing shows that patience can pay off: although captive orangutans are the most manipulative of all nonhuman species, for years no one saw wild orangutans regularly use tools until van Schaik reported habitual use of probes to get insects and scrapers to process hairy fruits. So it seems that orangutan material culture must be pursued in nature, too.

Gorillas present a curious challenge for cultural primatology. The only field study so far to yield detailed behavioral data on the species is of the most anomalous population, the highland gorillas of the Virunga volcanoes, made famous by Dian Fossey. This tiny, relict, high-altitude population bears little resemblance to the other 99 percent of the species, which lives farther west in Africa in lowland rainforests. Thus, the unusual diet of mountain gorillas reflects the alpine habitat; most species of social insects eaten "down below" are absent at 3,000+ meters! Similarly, mountain gorillas more often build their beds of vegetation on the ground than do their lowland counterparts, for the simple reason that nesting trees are scarce at such high elevations.

Little is yet known of the social life of lowland gorillas in nature, but differences in diet have emerged across populations, even in the same country. In Gabon, Caroline Tutin and Michel Fernandez compared insect-eating at Lopé and Belinga. At Lopé, gorillas eat mound-building termites but not weaver ants; at Belinga, gorillas eat weaver ants but not termites. Both types of prey are common at both places.

Of the 54 genera of primates, only a few have received much attention from cultural primatologists. A comprehensive survey of published scientific papers over 12 years, using index terms such as culture, tradition, and imitation, showed that 80 percent of the articles were on only five genera: *Cebus, Gorilla, Macaca, Pan,* and *Pongo*. Of these, over half were on one species, the chimpanzee, which justifies its treatment in a separate section.[3]

Chimpanzee Culture

Chimpanzees have been studied in captivity since Köhler's research in 1913–1917 on Tenerife, and in nature since Henry Nissen's trial effort in 1930 in Guinea. After Köhler, scores of laboratories have kept these apes for psychobiological investigation, although most have provided only impoverished living conditions that probably depressed the apes' intellectual performance. Enlightened zoos have done a better job, as best exemplified by Arnhem in the Netherlands; there the brothers van Hooff created a 1-hectare enclosure and an ape group of naturalistic makeup. For almost 30 years, Arnhem has provided the most useful data for comparison with the wild, from social hierarchies to tool use.

Modern fieldwork on chimpanzees began in the 1960s, with the trail-blazing research of Jane Goodall, Junichiro Itani, Adriaan Kortlandt, Toshisada Nishida, and Vernon Reynolds. Some of these pioneers remain active today: Goodall's project at Gombe in western Tanzania has now passed its fortieth year; Nishida's project in the Mahale Mountains, 140 kilometers south on Lake Tanganyika, is in its thirty-sixth year. Reynolds has resumed research at Budongo. Each site continues to yield new findings, as no study has yet surpassed the life span of the subjects of research: chimpanzees can live for more than 50 years!

Three other ongoing studies began in the 1970s: Bossou in Guinea, Kibale in Uganda, and Taï in Ivory Coast. Together with Gombe and Mahale, these constitute long-term studies in which known individuals are followed daily from dawn to dusk, thus affording a comprehensive (as opposed to selective) view of daily life. Altogether, at least 40 populations of wild chimpanzees have yielded ethnographic data, and at least 15 of these have been studied for more than a year. No other species of primate, except our own, offers such a rich array of diversity for analysis. By comparison, George Murdock's standard cross-cultured sample of independent human societies numbers only 43 for Africa.

The easiest sort of ape culture to study is material: apes leave recognizable artifacts, so even unhabituated populations can provide data for archaeological primatologists. Furthermore, all populations studied for at least an annual cycle are known to make and to use tools, so that elementary technology is a chimpanzee universal. (This fact suggests that, like us, the apes *depend* on tools.) Each group of chimpanzees has its characteristic tool kit, made mostly from vegetation, that serves for subsistence, defense, self-maintenance, and social relations.

Some items of material culture are virtually standardized, such as the sleeping platforms (also known as nests or beds), that every weaned individual makes every night. These shelters are constructed of interwoven fresh branches, usually in the canopy of leafy trees or vines. From Senegal to Uganda, the simple structures are remarkably similar and essentially interchangeable. They vary by size (according to the body weight of the maker) and species of raw materials used (according to the type of forest), but these are only minor variations on a constant theme. A Mahale chimpanzee would sleep comfortably in an Assirik chimpanzee's nest, even if 5,300 kilometers away from home!

In contrast, some material culture is impressively plastic (Figure 9.2). A chimpanzee may crumple a handful of leaves to make a drinking sponge, or use a single leaf to wipe her bottom, to probe for underground termites to eat, or even groom it as a prompt to elicit social grooming from another (Figure 9.3), or orally clip it to bits as a signal in courtship. Conversely, a

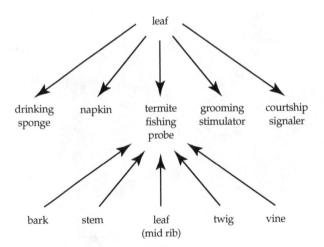

Figure 9.2 Double flexibility of chimpanzee tool use. One type of raw material (leaf) can be used in many ways, while many kinds of raw material can be used to make one type of tool (termite fishing probe).

Figure 9.3 At Gombe, a juvenile female chimpanzee uses normal grooming patterns to groom a leaf, while an attentive male infant observes. (Photo by W. C. McGrew.)

Figure 9.4 At Gombe, three adult chimpanzees use tools to fish for termites from a termite mound. (Photo by L. F. Marchant.)

probe for fishing termites from their mound (Figure 9.4) may be made from many kinds of vegetation, so long as it is straight and limber: bark, stem, twig, or vine, in addition to the midribs of a compound leaf.

What of culture that is not material, in order to satisfy the final criterion in the table? Social grooming is the glue of primate life, for it serves for more than hygiene: primates groom one another (Figure 9.5) to cement bonds, to reconcile conflicts, to solicit copulations, to calm distress, and (apparently) to pass the time. All chimpanzees show social grooming that is species typical, for example, two individuals may groom each other simultaneously, unlike monkeys or prosimians, who almost always take turns. At the same time, different populations show variants or nuances on a common behavioral pattern. These are local customs.

A small number of chimpanzee populations show the grooming hand-clasp, in which two grooming partners sit facing each other and extend one arm fully overhead and clasp hands, creating a sort of A-frame that exposes their underarms (Figure 9.6). This occurs every day at Mahale, but over decades has never been seen in near-neighbors at Gombe, although otherwise the basic grooming of the two populations is similar. A captive group of chimpanzees near Atlanta has also invented the grooming hand-clasp, and the progress of the custom is being monitored.

Occasionally, a variant of social grooming is unique to a population. Only the chimpanzees of Mahale have yet been reported to show the social scratch. All chimpanzees everywhere scratch themselves in a characteristic way by raking their fingertips over their trunk and limbs, but only Mahale's apes deploy this self-maintenance pattern to the social context. In the midst of a bout of social grooming, the groomer scratches the recipient for two or three strokes, then resumes normal grooming. What is the function? Who knows, any more than we know why some human beings cut their hair, and others do not?

It is evident that much of chimpanzee behavior fulfills the general principles for culture set out above. From others in their community these apes learn abiding acts that reflect the culture's unique, collective identity, embedded in the species' typical repertoire (or ethogram). So far as the eight criteria of the table are concerned, some are easier to check off than others. The most difficult is innovation. A field primatologist is fortunate indeed to be present for such a rare event, as is a cultural anthropologist studying a traditional, preliterate human society. But occasionally it happens. For example, at Bossou, after 20 years of negative data, Sugiyama's students found the chimpanzees using sticks to dip algae out of forest pools. This

Figure 9.5 At Gombe, a juvenile female chimpanzee grooms an adult female's back, probably seeking access to the infant held ventrally by the mother. (Photo by C. E. G. Tutin.)

Figure 9.6 At Mahale, two adult male chimpanzees engage in the grooming hand-clasp. Note the use of the thumb to groom the other's underarm. (Photo by W. C. McGrew.)

new technique and food stuff had never before been seen in other wild chimpanzees.

Some material culture of chimpanzees is directly relevant to the evolutionary origins of human culture. When apes use stone hammers and anvils to crack open nuts, these lithic artifacts survive to be recovered and analyzed by archaeologists, just like the tools of our extinct ancestors. Moreover, repeated use of these percussive implements leaves characteristic patterns of wear. Hammers and anvils used by chimpanzees at Taï and Bossou look remarkably like Oldowan tools recovered from such famous sites of early hominid activity as Olduvai Gorge in Tanzania. Paleoanthropologists are forced to broaden their interpretations of the function of these early artifacts: perhaps instead of being used by male hunters to crack open bones in order to extract marrow, they were used by female gatherers to crack open nuts to extract their "meat."

Only a few populations of chimpanzees are known to use hammers and anvils. Although the bulk of the fieldwork has been done in eastern and central Africa, the custom is found only in West Africa. Painstaking survey work shows that such nut-cracking occurs only west of a major river in Ivory Coast, suggesting that this technology was invented somewhere in the region of present-day Guinea/Ivory Coast/Liberia/Sierra Leone. Its diffusion to the east was apparently halted by the water barrier, for apes cannot swim.

Ecological research farther east at Lopé Reserve in Gabon, in central Africa, shows that all the necessary prerequisites for nut-cracking are available there to the chimpanzees: nuts, hammers, anvils. Also, the Lopé chimpanzees are not stupid; they use other, nonlithic tools. Nothing prevents those apes from nut-cracking except lack of cultural knowledge. Until some ape at Lopé independently invents the practice, or some immigrant introduces it (having somehow overcome the riverine obstacle to diffusion), the chimpanzees to the east remain in the non–Stone Age.

In summary, it may be that what chimpanzees do is indistinguishable from what humans do, in terms of operational, testable, behavioral criteria for culture. Of course, there are manifold differences. In African forests, some humans set snares to catch mammals to eat, but no chimpanzees do. In the same forests, some chimpanzees run down mammals in the upper canopy to eat, but no humans do. However, what chimpanzees think or feel or believe is another set of issues, subject to much more controversy.[4]

Critique

It is unlikely that any primatologist will ever interview a community of wild primates in either their language or ours. Thus, verbal self-report as a source of data is unavailable, and we lack the chief means of knowing what goes on in their minds. Must we then deny them culture, because our methods thus far are inadequate? Or must we remain agnostic, forced to categorize the question of cultural primatology as unknowable? I think not.

First of all, there are benefits as well as costs to such constrained methods. After all, primate informants will not tell us bold-faced lies, by omission or commission. Second, we can learn much just by watching and listening, even if we do not engage our subjects. This means using ethological methods, as developed by the Nobel Prize–winning trio of Konrad Lorenz, Niko Tinbergen, and Karl von Frisch. Just as von Frisch deciphered the dance language of honey bees, so can we make probabilistic inferences about the acts of primates, based on stimulus-and-response sequences, on successful or failed outcomes, or on modified repetitions.

This sequence is easily demonstrable with fellow humans. Repeated viewing of archival, unedited videotape of, say, East African foraging people in daily life *is* informative, even if they are strangers speaking a language of which the viewer is ignorant. Making sense of what primates do invokes many of the same capacities; of course the resulting knowledge is incomplete, but it is more accurate with each new round of hypothesis posing and data collection.

Psychologists critical of cultural primatology focus on what they see as the absence in nonhumans of certain critical mechanisms of transmitting knowledge. Chief among these are the processes of imitation and teaching. In this special sense, imitation refers to a higher-order, representational mental process, not just to the mindless mimicry of mynah birds. Similarly, teaching refers to the intentional passing of information from a knowledgeable individual to a naive one, that is, active and targeted social tuition. The debate has now largely been defused, thanks to the ingenious experimentation in captivity of Andrew Whiten and the painstaking observations in nature of Richard Byrne. Apes in captivity imitate one another, and if those in nature cannot be proven to do so, it is because of Alice in Wonderland logic. The argument goes: To demonstrate imitation requires that all variables be controlled, and this calls for experimentation; but true experimentation in nature is impossible, for when nature is controlled, it becomes artificial and thus no longer nature. Therefore imitation can never be shown in nature.

To anthropologists, neither of these processes is very interesting, as what matters to them is not *how* knowledge is conceived and disseminated but *what* knowledge is shared. A common objection is that culture and language are inseparable, and so by definition a nonlinguistic creature could not be cultural. The logic is suspect, for language could just as readily depend on culture as vice versa, or both could be by-products of the general intelligence that coevolved with large brains. Until we can find a human society that lacks either language or culture, the question will remain unresolved. (Studies of human-infant development suggest that coevolution is most likely: in their first year of life, infants show few signs of language or culture but do show surprising abilities to reason.)

The standard line of argument is predicated on the assumption that nonhuman primates lack language, which is also questionable. Studies of captive chimpanzees show that a youngster can learn American Sign Language from other chimpanzees (that is, without any human intervention) and that these chimpanzees communicate with signs among themselves, even with no humans present. And studies of wild chimpanzees (and other primates) show that dialect differences exist in vocalizations across populations. The chimpanzees of Gombe, Kibale, and Mahale can be distinguished on the bases of tape recordings of the same type of call, such as a pant hoot. If French is spoken differently in Paris, Brussels, Montreal, and Louisiana, we deem these differences cultural. What are we to do with similar vocal diversity in primates? Until linguists are willing to go to the field and tackle directly the communication of primates in nature, the question of language remains open.

In a critical review, the primatologist Christophe Boesch and the psychologist Michael Tomasello focused on what they call a key feature of culture, the "ratchet effect." It is the accumulation of modifications in traditions over time, so that their utility is ever enhanced. The ratchet effect is hypothesized to be a difference between humans and other species, with human culture being advantaged by its existence.

Still, the progressive functioning of the ratchet has already been seen in the Japanese macaques of Koshima. At the outset, in 1958, the monkeys carried handfuls of sand and wheat grains to the sea and cast this mixture into the water. The sand sank, the cereal floated, and the food could be easily skimmed from the surface. When dominant individuals began to usurp the proceeds from sluicing subordinates, those of low rank devised counter measures, such as keeping the sand-grain mix in their fists while sluicing, thus never letting the food out of their control. Later some individuals even began to dig small pools near the shore, in order to practice solitary, more secure flotation. Over 25 years the monkeys developed eight such new techniques, thereby ratcheting up the basic, initial procedure. Similar improvements occurred in the techniques of sweet-potato washing.

Evolution of Culture

Natural selection for cultural capacity is just as amenable to analysis by Darwinian evolutionary theory as is that for any sensory or motor capacity. The ultimate test is whether or not behavioral or cognitive traits influence the relative fitness of their bearers. Traits that increase inclusive fitness will increase in relative frequency. That culture is inherently social (rather than physiological or morphological) should make no difference in such analyses, as the survival values of many other social phenomena are well known (imprinting, attachment, dominance rank, and so on). Nor does the malleability of culture present an obstacle to selection; rigid reflexes may be easier to program than labile customs, but plasticity can be more beneficial, especially in changing environments. Quickly acquired cultural knowledge is a powerful tool.

The more complicated the cognitive processes involved in social learning, the more costly these are likely to be in terms of risk, if not of time and energy. If more complex cognition requires bigger brains, and if bigger brains are expensive organs to fuel, then energy is a constraint. Regarding the saving of time, whether, say, matched dependent learning is more time-consuming than observational conditioning is not clear; all that is certain is that one-trial learning is the most economical.

Risk has been underemphasized as a cost of social learning, and so acculturation too is risky. Neophytes may be hindered by becoming attached to inferior models, or customs that sufficed for the parents' generation may be outdated for their offspring. Would it be better to learn from peers? But peers are likely to be competitors, who may "hold back" as models or, even worse, mislead. Better to stick to relatives, and trust that blood is thicker than water—but how to choose? Just because Brother A is closer kin than Auntie B does not make him the better model to emulate. Closer relatives may be more patient or forgiving with a pupil, but why should they be more proficient? The potential costs of such hazards to a learner are immense, even in passive social learning, in which the model merely goes about its business of foraging, fighting, or the like while the novice pays attention.

If all of these costs (energy, time, risk) rise with the plasticity of behavior, then the more complex a process, the more potentially costly it is. In terms of cost minimizing, an organism should seek to gain reliable, valid information as simply as possible, and so "gear up" to the next cognitive level only when forced to do so by stricter selection. As in most matters, simpler is cheaper is better.

Teaching then becomes the ultimate high-risk/high-payoff strategy of acquiring culture. Teaching is different, because for the first time more knowledgeable individuals adjust their actions (presumably at some cost to themselves) for the sake of instructing less knowledgeable subjects (who presumably benefit accordingly). What was, at an earlier evolutionary stage, one-way passive learning becomes two-way interactive teaching. Instead of only inadvertently setting an example, tutors can be selfish or spiteful, as well as altruistic to or cooperative with their pupils. If pupils depend on mentors for knowledge, then they are vulnerable to manipulative tuition by mentors who act out of self-interest. In evolutionary terms, the role of teacher should be no more self-sacrificing than any other role, be it that of mate, leader, or even parent.

The risks of allowing oneself to be taught are evident. Teachers may fail to teach optimally, especially if the knowledge imparted makes pupils better competitors, as when resources are scarce. Teachers may teach inefficiently or even misinform; for example, if a more learned pupil might usurp the teacher's status. Such is the risk of apprenticeship. Even in cooperation, when both parties benefit, a teacher may ensure that a pupil's net benefit remains less than her own. Just as genetic parent-offspring conflict is inevitable in the social evolutionary framework of Robert Trivers, so too may be mimetic teacher-pupil conflict. Put another way, if second-order intentionality (I know *that* you know) is arguably a necessary prerequisite for teaching, it can be extended to exploitable knowledge of another's ignorance (I know *what* you know or don't know). In summary, being taught should be the last evolutionary resort for any cultural organism, to be used only if simpler, self-reliant means are not enough. It then comes as no surprise that traditional (nonliterate) human cultures seem to avoid teaching whenever possible. For instance, Barry Hewlett and Luigi Cavalli-Sforza showed that the 50 basic skills of Aka pygmy daily life were mostly acquired passively, without teaching.[5]

Addition of symbol-based communication—that is, language—to the equation accentuates some of the above processes, but does not replace them. Sociality is a necessary but not a sufficient condition for culture; we do not seek culture in schools of tadpoles. In contrast, language is not a necessary condition for culture, but it is a sufficient one. Chimpanzee culture (as outlined above) need not be linguistic, but if a person speaks German, one knows that she has been enculturated.

However, the addition of language transforms teaching. Linguistic communication frees knowledge transmission from the here and now of performance. Recorded language frees teaching even from the *presence* of the teacher, for the pupil can learn from artifacts in her absence. Language-using teachers can refer to yesterday's activities done elsewhere or can project to routes to be traveled tomorrow. Linguistic instruction can propose not-yet-needed solutions to not-yet-encountered problems, or reconsider past failures. Clearly, this is marvelous for culture—but it can also be used to turn everything upside down. Linguistic acculturation can also be used to confuse or deceive, as in bigotry and propaganda. Put another way, only voyeurs acquire honest (but likely limited) knowledge, while spectators must interpret as they observe; but pupils, whether listening or reading, must be ever vigilant. Culture does not come free.

In summary, ethnologists infer the cultural minds of their human informants from a combination of participant observation and conversation. From this melange of subjectivity and verbiage emerges an ethnography. Ethologists infer the cultural minds of their nonhuman subjects by striving for detached, objective observation without interaction; from this results an ethogram and data set.

Each approach has its pros and cons. Informants lie by work and deed and adapt to their ethnographers, and vice versa. Arguably, all observations made by participants are compromised by conflict of interest, yet refusal to take part may preclude being informed.

Ethology seems a sparse method by comparison: what you perceive is what you get. Thus, biased or distracted or incompetent observers produce flawed data. Primates may deceive their investigators, but only by deed, which may be more detectable than by word. All in all, cultural primatologists would seem to have the easier task, but time will tell.

Conclusion

Homo sapiens is not the only cultural species of the order Primates. If human culture emerged out of nonhuman nature, then we must seek the intermediate stages of this evolutionary transition if we are to understand more fully our cultural roots. To the traditional approaches of cultural anthropology and archaeology, we must add the wider perspectives of comparative psychology and behavioral ecology. Thus, cultural primatology is a genuinely new synthesis of theory and methods that crosses the disciplinary lines of social and natural science. But our efforts will only be as meaningful as our material: enriching the daily lives of captive primates and preserving the habitats of wild primates is the necessary foundation of productive cultural primatology.

Notes

1. Of Genes and Apes

1. When individuals of species in which inbreeding normally does not occur do mate with close relatives, they usually experience significant inbreeding depression, probably because recessive deleterious alleles are expressed in the homozygous form (see review by Pusey and Wolf, 1996). For example, Ralls, Ballou, and Templeton (1988), examining 40 captive populations of 38 species of mammals, found that the survival of offspring from parent-offspring and sibling matings was reduced on average by 33 percent. This figure is likely to be an underestimate of total inbreeding depression, because inbreeding also reduces fecundity and increases sterility, adult mortality, and susceptibility to disease.

2. Behavioral avoidance of close inbreeding is also common in other species of primates (Pusey, 1990b), and in animals in general (Pusey and Wolf, 1996).

3. My colleague Craig Packer and I explained the pattern of female philopatry and male dispersal, and the rarer cases of the opposite pattern, on the basis of the costs and benefits of dispersal to each sex (Pusey, 1987; Pusey and Packer, 1987; but see also Moore and Ali, 1984). Because female reproductive success depends mostly on the speed at which the females can produce offspring, they will benefit by staying in a familiar area in which they can feed efficiently and avoid predators, and possibly gain advantages from the presence of female kin with whom they can cooperate. On the other hand, males, whose reproductive success depends on the number of females they can inseminate, may benefit by moving around to gain access to more females. Also, because there tends to be greater competition between males for access to females than the reverse, males may sometimes benefit by going to groups in which there is less competition or a better female-to-male ratio. We argued that males only stay when there are unusual advantages to doing so, and then females are "forced" to leave to avoid inbreeding (see also Pusey, 1979). Along the same lines, Clutton-Brock (1989) showed that female mammals are generally more likely to disperse in species in which males remain in the social group longer than the age at which females reach sexual maturity.

4. Infanticide by female chimpanzees has only been observed at Gombe. In 1975 high-ranking female Passion and her adolescent daughter attacked the low-ranking female Gilka, and seized, killed, and ate her infant. Later they

snatched and ate Gilka's next infant and were suspected to have killed at least six other infants during the same period (Goodall, 1977, 1986). At first it was thought that Passion and Pom's behavior was aberrant and pathological. Still, another high-ranking female and her adult daughter (with, in one case, a third unrelated female) have now been observed to make determined but unsuccessful attempts to seize infants on two separate occasions, and a group of females in an adjacent community were observed eating an infant (Pusey et al., 1997).

5. Extraction and amplification of nuclear DNA from feces and hair to produce accurate genotypes is technically difficult because the quantities of DNA are often small (Taberlet et al., 1999). Each gene occurs in two copies in an individual's body, one from the mother and one from the father. These copies (alleles) may be the same (homozygous) or different (heterozygous). Sometimes only one allele is amplified and detected in one sample (allelic dropout). So to detect a heterozygote reliably one needs to repeat the procedure a number of times. In addition, detecting differences between the alleles is difficult if they are similar but not the same in size. Whereas early studies used genes whose alleles could differ in length by just two base pairs, later studies limit analysis to genes whose alleles differ by at least four base pairs, so that scoring errors are reduced. Cases of allelic dropout or misscoring can influence calculations of relatedness and can also lead to false exclusion of males as fathers.

6. Roger Short (1979) has related the astonishing variation in sexual dimorphism and size of the genitalia in the great apes to variation in social structure and mating patterns. Unlike chimpanzees, gorillas live in small groups with only one or sometimes two adult males; of these, one is strongly dominant over the other. Intense competition occurs between males for initial acquisition of females, but once a male has established a group, he is usually the only male to mate with the females. As a consequence, males are extremely large compared to females, reflecting the strong sexual selection for fighting ability. But females do not advertise receptivity, mating is infrequent and low key, and the male has a tiny penis and testes relative to his body size.

Chimpanzees show more moderate sexual dimorphism in body size and weaponry, but have relatively enormous testes and a long penis that has probably evolved to reach down the large vaginal swelling and circumnavigate any sperm plugs from other males. A strong relationship between size of testes and females mating with multiple males has been found among primates in general (Harcourt et al., 1981). Human testes are relatively smaller than those of chimpanzees. Short suggests that human testes size and physiology are adapted to low but continuous rates of copulation. Frequent ejaculation in humans leads to a marked depletion of sperm and an ensuing short-term decrease in fertility. The human penis is much thicker than that of any ape. Interpretations of the significance vary from Short's view that the penis is a visual and probably tactile display to the female, to Jared Diamond's view that it evolved as a display between males.

2. Apes from Venus

1. Fossil evidence from shortly after the divergence between our direct ancestors and the African apes suggests that the dwelling place of the earliest hominids was still to some degree forest habitats. Most likely, the transition to bipedal locomotion in hominid evolution was gradual: the "savanna hypothesis," which assumes a sudden and early adaptation to open habitats, is under increasing scrutiny (Shreeve, 1996).

For example, Clarke and Tobias (1995) unearthed four small bones at Sterkfontein that formed part of the left foot of a human ancestor estimated to have lived at least 3 million years ago. The foot had the weight-bearing heel adapted for bipedal locomotion, but also an apelike toe that could still grasp. This australopithecus fossil, baptized Little Foot, may stem from a time when our ancestors still needed to get into the trees to collect fruits or escape from predators.

The controversy surrounding this fossil has made scientists take another look at the bonobo, which has the longest legs of all four extant apes (Zihlman et al., 1978; Zihlman, 1984), is a competent bipedal walker, but misses the specific ankle that makes Little Foot, and not the bonobo, one of our forebears. Documenting bonobo locomotion at Lomako Forest, Susman (1984a) saw parallels with early hominids. Not only bipedality, but also other aspects of hominization such as tool-making and cooperative hunting, may have taken place in the forest before our ancestors began putting those skills to use on the plains. Boesch and Boesch (1994) defend this thesis based on observations of forest-living chimpanzees. Roosevelt (in press) reviews evidence that moist tropical forest habitats, such as those of the Congo Basin—currently shared by bonobos and human pygmies—may have played a larger role in human evolution than the savanna. At the same time, she questions the killer ape myth, arguing that there is little or no fossil evidence from before the agricultural revolution for habitual intercommunity violence in our lineage. Roosevelt thus postulates geographic, ecological, and behavioral similarities between bonobos and protohominids.

2. Nimchinsky et al. (1999) report large, spindle-shaped cells in the anterior cingulate cortex of great apes and humans. These cells do not occur in gibbons or monkeys, or any other mammalian taxa. The restricted distribution, combined with the fact that in humans these neurons are affected by Alzheimer's, could mean that the cells are involved in higher cognitive processes. Of the four great apes, bonobos show the most humanlike clustering of these neurons.

3. The debate raged especially between 1966, after the English translation of Lorenz's *On Aggression,* and the appearance in 1978 of E. O. Wilson's *On Human Nature.* Leading opponents of Lorenz included ethologist Hinde (1970), anthropologist Montagu (1976), and a number of social psychologists (reviewed in Berkowitz, 1993). In those days all issues pertaining to human evolution were placed in the context of the origins of aggression. The debate continued with a book by a Montagu follower who claimed that wild chimpanzees undisturbed by human observers are completely peaceful and egalitarian (Power, 1991), and

the *Seville Statement on Violence,* which reads like a belated point-by-point rejection of Lorenzian views (Adams et al., 1986; criticized by Fox, 1988, and de Waal, 1992a). Even though today's views are better informed and based on more sophisticated evolutionary thinking, old themes have reemerged in *Demonic Males* (Wrangham and Peterson, 1996) and *The Dark Side of Man* (Ghiglieri, 1999). Other scientists in a variety of disciplines, having abandoned the obsession with aggression, are exploring how evolution has equipped our own species as well as many others with powerful mechanisms to manage and resolve conflict (reviewed in Aureli and de Waal, 2000).

4. Even if female chimpanzees normally do not show the degree of bonding seen in bonobos, we need to realize that female-female relations are probably the most variable element of chimpanzee society. These relations range from close and mutually supportive in captive colonies to rather loose and indifferent in some wild populations (e.g., Goodall, 1986). But also in nature, remarkable variability exists. Female bonding appears to characterize a small chimpanzee population "trapped" by agricultural encroachment in a forest of approximately 6 square kilometers on top of a mountain in Bossou. Sugiyama (1984) often saw the majority of individuals in this forest travel together in a single party, and he measured relatively high rates of female-female grooming. Similarly, female chimpanzees have been said to be highly sociable in the Taï Forest (Boesch, 1991). Even if it now appears that the difference between Taï and other sites is not so great as originally claimed (Doran, 1997), it nevertheless remains useful to think in terms of *adaptive potentials* (de Waal, 1994). In other words, chimpanzees have a potential for female bonding. In most parts of their range this potential goes unrealized because of ecological pressures to disperse, but under certain conditions the female relations of chimpanzees can become rather bonobolike. This tendency is particularly true in captivity, where dispersal is impossible. On the other hand, overlap with bonobo social organization is incomplete: captive chimpanzee females never reach the level of social integration seen in their bonobo counterparts, nor do their occasional alliances translate into the same high social status (Parish, 1993, 1996).

5. See Stanford (1998) for this discussion, which includes replies by Kano, Fruth, de Waal, and others. Because of the reluctance of scientists unfamiliar with bonobos to accept certain species characteristics, Parish and de Waal (2000) have recently drawn up a list of observed behavioral tendencies (females chase males, females associate preferentially with each other, bonobos are relatively peaceful) and the unparsimonious explanations proposed by skeptics (males are chivalrous, females merely tolerate one another, violence does occur but has not yet been observed).

6. It is highly unlikely that nonhuman animals recognize the connection between sex and reproduction. When we say that paternity is "obscured," we refer to the impossibility that evolution design simple rules of behavior to make males treat their offspring differentially. It has nothing to do with males knowing which infants they have fathered—even though for convenience we may

phrase it this way—but rather with behavioral rules that foster special relationships between males and potential offspring via the male's past and present contacts with the mother. Infanticide has been documented in a great variety of animals (Parmigiani and vom Saal, 1994) and its explanation is still under debate (see *Evolutionary Anthropology*, 1995). In contrast to the repeated reports in chimpanzees (e.g., Nishida and Kawanaka, 1985; Goodall, 1986), the behavior has thus far never been documented in bonobos. If it is truly absent, that fact may help explain the relatively high survival rate of bonobo infants (Furuichi et al., 1998).

7. Sex-for-food exchanges between the sexes may be a vestige of an evolutionary past in which bonobo females had not yet attained dominance over males. Chimpanzee males dominate females, and this tactic characterizes females of all ages (see Chapter 4 and Yerkes, 1941). In bonobos, however, the tactic is typical of young females. Fully adult females have no need of male tolerance: they simply claim access to food.

3. Beyond the Apes

1. It is important to point out the anthropocentric emphasis on apes and semiterrestrial monkeys, which reflects traditional anthropological concerns with finding the best candidates for comparison with humans (Richard, 1981). Primatologists trained in other disciplines have tended to be less concerned with identifying suitable referential models for human behavioral evolution, and more interested instead in the comparative ecological and social diversity among primates and between primates and other animals (Tooby and DeVore, 1987; Rowell, 1993).

2. Some of the first phylogenetic analyses of primate behavior involved hominoid (Wrangham, 1987) and platyrrhine comparisons (Strier, 1990, 1994; Garber, 1994), but Di Fiore and Rendall (1994) were the first to demonstrate the behavioral distinctiveness of cercopithecines using contemporary cladistical methods. Macaques and baboons are not "typical" primates. Their analysis was particularly persuasive in linking female matrilineal residence and other attributes of cercopithecine female bonding into a social combination possessed by few other anthropoid primates.

3. Life-history variables that affect reproductive rates, including female age at first birth and interbirth intervals, also appear to be phylogenetically conservative traits. With the exception of the callitrichids (marmosets and tamarins), most species of platyrrhine monkeys are more like apes than Old World monkeys in terms of their life histories and reproductive rates (Ross, 1991). Parallels between platyrrhine and hominoid life histories and dispersal patterns suggest that reproductive rates and dispersal patterns may be linked (Strier, 1996, 1999a). For example, slow reproductive rates will constrain the competitive options available to males, and could favor cooperation over competition to monitor females and their offspring. Conversely, the greater reproductive rates rela-

tive to body size among Old World monkeys, by making matrilineal allies more important to females, might underlie the taxonomic differences in their dispersal patterns when compared to other primates.

4. For more details on these studies, see Palombit (1994, 1995) and Reichard and Sommer (1997) for gibbons; Robbins (1995, 1996) and Watts (1996) for mountain gorillas; Mitchell (1994) for Peruvian squirrel monkeys; Sussman (1992) for lemurs; Cheney and Seyfarth (1983) for vervet monkeys; Garber (1994) for callitrichids; and Strier (1999a and b) for platyrrhines. For more comparative perspectives, see Moore (1984, 1992) and van Hooff and van Schaik (1992, 1994) on the variants of female bonding and male bonding across primates.

5. The longstanding perception that bonobos and chimpanzees are different from one another has been challenged by Craig Stanford (1998a and b). His argument has inspired lively debate, including a series of commentaries to which he has responded. Stanford's goal was to stimulate closer consideration of the assumptions we make regarding apes and what they imply about our own behavioral evolution. Nonetheless, respondents were largely consistent in maintaining that the apparent behavioral differences between these species are real. Of particular interest here is that while male bonobos and chimpanzees may copulate at comparable rates, when the number of available females and the rates of sexual interactions between females are taken into account, bonobo sexuality still stands out (Kano, 1996; Takahata et al., 1996).

6. Further insights into female reproductive strategies, originally described by Hrdy (1981), can be found in Smuts (1985), Small (1988, 1989, 1992), Smuts and Smuts (1993), and Hrdy et al. (1995). Alternative interpretations of selection pressures favoring female strategies to confuse paternity and thereby reduce the risks of infanticide can be found in Bartlett et al. (1993) and Sussman et al. (1995).

4. The Ape's Gift

1. For more information on chimpanzee hunting behavior and its relevance to human evolutionary research, see Goodall (1986), Boesch and Boesch (1989), Boesch (1994), Stanford et al. (1994), Stanford (1996, 1998a), and Uehera (1997).

2. While 15 percent mortality due to predation has been recorded for other species of mammals, it must be remembered that this figure represents predation by chimpanzees only and does not include death at the hands of other predators (leopards and eagles occur at Gombe and are known predators of monkeys) or mortality caused by disease, infanticide, or other factors. A 35 percent mortality rate, if it occurred every year, would mean that the red colobus population would almost certainly be in sharp decline. The explanation here is that the average annual mortality in the colobus population due to chimpanzee predation, taken over the past decade, is about 20 percent per year.

To learn whether chimpanzee predation has the potential to regulate the size of the colobus population, I compared the intensity of hunting by chimpanzees with the size of red colobus groups in each of the valleys of the chimpanzees' hunting range. The central valley of their range (the so-called core area) is Kakombe Valley; the Kasakela chimpanzees made about one-third of all their hunts there over the past decade. Chimpanzee use of the more peripheral valleys is much less frequent, and their frequency of hunting there is lower also. Only about 3 percent of all hunts took place at the northern and southern edges of their range. The size of red colobus groups also varied over the area of the chimpanzees' hunting range. In the core area, red colobus groups averaged only 19 animals, little more than half the average of about 34 at the outer boundaries. In other words, colobus groups are small where they are hunted frequently, and larger where hunting is infrequent. The reason for this size difference is largely the difference between core area and peripheral groups in the percentage of immature colobus. In the core area, only 17 percent of each group was infants and juveniles, while fully 40 percent of peripheral groups were immature (Stanford, 1998a).

3. The fossil record for meat-eating is unambiguous as to its origin about 2.5 million years ago, but the hunting/scavenging debate continues. For in-depth discussions see Isaac (1978), Shipman (1986), Blumenschine (1987), and Bunn and Ezzo (1993).

4. See Hill and Hawkes (1983) and Hawkes (1991).

5. For more on tactical deception and social intelligence, see Chapter 6 and Byrne and Whiten (1988), Dunbar (1992b), and Byrne (1996b).

5. Out of the *Pan*, Into the Fire

1. In this chapter I follow current practice by considering hominids, or the family Hominidae, to include both hominins and the African great apes. The hominins are all the species of the tribe Hominini, that is, the bipedal hominids. Wood and Collard (1999) list five genera: *Ardipithecus, Australopithecus, Paranthropus, Praeanthropus* (all of which are australopiths), and *Homo*. The only known species of *Praeanthropus* is *P. africanus*, which has traditionally been named *Australopithecus afarensis*. I include it here as a member of the genus *Australopithecus*. The second tribe of Hominidae is the Panini, which includes only the three African great apes: chimpanzee *Pan troglodytes*, bonobo *Pan paniscus*, and gorilla *Gorilla gorilla*.

2. In this discussion of ancestry, I deliberately mention "chimpanzees" more than "bonobos." That focus comes from our reconstruction of whether the last common ancestor of chimpanzees and bonobos was more like chimpanzees or more like bonobos. We follow a simple procedure: we ask if gorillas are more similar to chimpanzees or to bonobos.

Overall, the answer is clear. In many ways gorillas are merely a large version of chimpanzees—for example, in their cranial and postcranial anatomy. Bonobos, on the other hand, are more gracile, smaller headed, and relatively juvenilized and sexualized compared to the other apes. Gorillas are thus more similar to chimpanzees than to bonobos, which are best thought of as a specialized offshoot of the gorilla-chimpanzee line.

What, then, are we to make of traits shared by bonobos and humans, but not by other apes? Examples are a tendency toward ventro-ventral copulation, a trend toward female-female bonding, and canines that differ little in size between females and males. Convergence appears to be responsible for these similarities. For example, reduction in sexual dimorphism of canine teeth is known to have developed more in later than in earlier australopiths.

Chimpanzees are therefore the more relevant model of *Pan prior*, our prehominin ancestor. Note that this conclusion is reached without considering a single behavioral similarity between chimpanzees and humans. It comes from comparing physical traits. It does give support, however, to the idea that behavioral traits that are shared by chimpanzees and humans, such as lethal intergroup aggression, may have a common ancestry that extends back to *Pan prior*. See Chapter 2 for a different view.

3. Peter Wheeler has made a plausible case for the major relevance of heat stress. By adopting an upright stance, he calculated, a sweating ape can reduce her incident radiation sufficiently to halve her water requirements. Chimpanzees prefer to avoid the midday sun, so it is reasonable to imagine that the reduction of diurnal heat load would have been a significant selective force for an ape whose original habitat was the forest, now forced to travel in more open habitats. Yet the trees that were so important to early hominins must often have provided shade. Other suggestions for the benefits of bipedalism include improved carrying ability (for instance, if it was necessary to carry digging-sticks or USOs), more efficient locomotion (allowing longer day-ranges), and the avoidance of transition costs between bipedal feeding postures and quadrupedal travel. None of these ideas is convincing yet (Chaplin et al., 1994).

4. The importance of body size for grouping pattern is apparent when analyzing the ultimate reasons why groups split up. In gorilla groups, individual females are attracted to a mature male who protects them against threats such as infanticide and predation. But chimpanzees feed and travel in mixed-sex parties only when conditions are favorable. At other times they split up and travel alone. Why do chimpanzee groups break up more often than gorilla groups?

Species differences in food choice provide the answer. When fruits are scarce, gorillas abandon the search for fruits. Instead, they settle into an easy dependence on the terrestrial herbs that are abundant in most gorilla habitats. As a result of this shift in food choice, individuals within a group have little need of competition with one another and can reside in stable groups.

Chimpanzees, by contrast, never give up searching for fruits, even when these high-quality items are rare. They experience "scramble competition," which is one of the costs of traveling with others. Scramble competition occurs when fruit trees contain fewer meals than a party requires. At such times each individual leaves the tree unsatisfied. To quell their hunger, all must travel farther than if they were alone, to find additional fruit trees. This extra travel, a key cost of scramble competition, can be reduced very simply. Individuals can travel in small parties, alone if need be, so that they can fill their bellies at each tree they visit. Both chimpanzees and bonobos follow this strategy. The benefit of reducing the cost of scramble competition during periods of fruit scarcity explains tidily the fission-fusion foraging arrangement of these species.

Note that the key difference between gorillas, on the one hand, and chimpanzees and bonobos, on the other, is their digestive response to fruit scarcity. Gorillas abandon the search for fruits, shifting instead to easily found herbs; the smaller chimpanzees and bonobos continue to search for fruits. So the critical question in deciding whether the woodland apes had stable groups or a fission-fusion system is whether they ever abandoned the search for high-quality foods.

5. Humans are the only primates with pair bonds among males and females who forage separately, a trick achieved by using language to monitor each other's activities. Since deception is possible, pair bonds are easily broken in a system that relies on language for mate-guarding. Thus, human pair bonds are rarely as exclusive as those of nonhuman primates.

6. Aggressive intergroup encounters are ubiquitous among group-living primates. Yet chimpanzees and humans are the only species of primates in which coalitions of males are known to kill rivals in intergroup encounters. These killings are apparently facilitated by a fission-fusion pattern of grouping. It is a system which means that occasionally a large party encounters a lone individual; by collaborating in their attack, the aggressors can kill with impunity.

Among mammals other than primates, coalitionary killing has been described in wolves (*Canis lupus*) and reported anecdotally for several other carnivores. The ultimate benefit of killing rivals appears to be a weakening of the territorial power of the neighboring community. This weakening is expected to occur in any terrestrial species that forms groups, has hostile relationships with neighbors, and in which parties of three or more sometimes meet loners (Wrangham, 1999). These comparisons suggest that if the woodland apes foraged in a fission-fusion system comparable to that of the chimpanzees (as it seems likely they did), they would have occasionally raided and killed their neighbors.

If so, the threat of violence near their borders would have been an important influence on behavior. For instance, woodland apes would probably have avoided border areas unless they could visit them in parties of several males. They would have patrolled their territorial borders, and they would have placed a premium on males supporting one another in interactions against neighbors.

6. Social and Technical Forms of Primate Intelligence

1. In the United States, psychology was dominated by behaviorism for the first half of the twentieth century; various schools existed, the most extreme be-

ing the radical behaviorism of B. F. Skinner. The cardinal features of behaviorism (Broadbent, 1961) are a commitment to basing the study of the mind on observable behavior (and indeed, a behaviorist would not use the word *mind* at all) rather than introspection or pure theory; a belief that in essence the minds of all species are the same, so that controlled laboratory study of the white rat or the pigeon would give all the data needed to understand the human mind (where humans were studied, the tasks were artificially simplified by the use of nonsense materials); and a ban on the use of unobservable intervening variables in explanations, so that "theory" became little more than redescribed data (Hodos and Campbell, 1969).

Of course, not all American behaviorists adhered rigidly to these tenets, and in Europe the situation was very different. European psychologists (notably Piaget in Switzerland, Bartlett in England, and de Groot in the Netherlands) continued to study human thought and memory as evidenced in complicated, everyday tasks, using elaborate and unobservable mental models to explain their data, and generally ignoring animals. Animal behavior was productively studied in the new field of ethology (Lorenz, von Frisch, and Tinbergen were jointly awarded the Nobel Prize for their work, and many others made major contributions). Like behaviorists, ethologists studied and measured behavior, but they began with natural behavior in the wild, only later setting up experiments in the field or laboratory to test hypotheses derived from their observations. In complete contrast to behaviorism, ethology treats each species as potentially different, relating the differences to evolutionary adaptation to particular environments and social demands, and freely postulates mental structures to explain observed behavior. (Although this field was pioneered chiefly in Germany, the Netherlands, and Austria, World War II led to Britain's inheritance of its legacy when Tinbergen moved to England.) By 1950 neither psychology nor animal behavior was a priority in a Europe shattered by years of war and adjusting to new cold war realities; most psychology was performed in the United States in the behaviorist tradition, and most animal behavior was studied in Skinner boxes as part of this enterprise.

The transformation of psychology to its modern form has been sufficiently dramatic to warrant the term *cognitive revolution*. It was largely accomplished by 1970, although the development of links with neuroscience and an ethology revitalized by sociobiology are continuing today.

In complete contrast to the ideals of behaviorism, the "mental models" proposed as explanations in cognitive science are often complex, and they are never directly observable. Some cognitive psychologists explicitly model their hypotheses on computers, but all are concerned that their models should be "adequate"—that they would work if machine implemented. Few psychology departments retain white-rat colonies for the study of behavior. Only the behaviorist prescription of basing theory on observable and measurable behavior remains.

2. Social complexity in the lives of monkeys and apes has been increasingly documented in the many field studies of primates carried out in the second half

of the twentieth century. The result was a number of suggestions that the evolution of intelligence is related to social challenges. As long ago as the 1950s, Michael Chance and Allan Mead (1953) pointed to the extended receptivity of female primates and the conflict situations thereby set up for males. They argued that taking into account the movements of both the female and a competing male during maneuvering posed a peculiarly difficult problem. The complexity of solving this problem, they postulated, led to an increase in the size of the neocortex in primates (they did not explicitly mention intelligence). In the early 1950s, sexual conflict was still seen as the basis of primate society (Zuckerman, 1932). Subsequently this exclusive emphasis on male-male conflict as demanding intelligence has found little support, and thus reducing the impact of Chance and Mead's speculations.

Alison Jolly (1966), having studied lemurs in Madagascar, noted that they lacked the intelligence of monkeys yet lived in similar-sized groups. She realized that this pattern was inconsistent with the idea—popular at the time—that monkey-level intelligence is necessary for long-term group living. She suggested instead that group living, arising without great *need* of intelligence, would subsequently tend to *select* for intelligence.

Nicholas Humphrey (1976) argued that monkeys and apes appear to have "surplus" intelligence for their everyday wants of feeding and ranging. Since evolution is unlikely to select for surplus capacity, he suggested that primate (and human) intelligence is an adaptation to social problem-solving. Group living inevitably causes competition among individuals, yet overall it must be beneficial to each member or it would not occur. For each individual primate, this situation sets up an environment favoring the use of social manipulation to achieve individual benefits at the expense of other group members, but without causing such disruption that the individual's membership in the group is put in jeopardy. Particularly useful to this end would be manipulations in which the losers are unaware of their loss, as in some kinds of deception, or in which there are compensatory gains, as in some kinds of cooperation. Intelligence is thereby favored as a trait, and since this selective pressure applies to all group members, an evolutionary arms race is set up, leading to spiraling increases in intelligence. Humphrey suggested that the resulting intelligence in social primates would likely be of a sort particularly suited to social problems and not well tested by the gadgetry of psychologists' laboratories, thereby explaining the many failures to find differences in intelligence between animals (see Warren, 1973; Macphail, 1982).

These arguments are somewhat different from each other and apparently were derived entirely independently; yet they share the feature that social complexity is given a causal role in the evolution of intelligence (see also Kummer, 1982). Andrew Whiten and I brought the theories together in the late 1980s and later (1997) published a volume that included some of the most striking data on primate social complexity. The book's title, *Machiavellian Intelligence*, was inspired by Frans de Waal's (1998 [1982]) explicit comparison between chimpanzee social strategies and some of the advice offered four centuries earlier by

Niccolò Machiavelli. Machiavellian intelligence seemed an appropriate metaphor in that so many features of primate behavior appeared to be cooperative and helpful, yet evidently resulted from natural selection's maximizing the inclusive fitness of certain individuals relative to others.

The extent to which this theory has been successful in accounting for the variation in cognition among primates, and for the origins of intelligence in the human line, is evaluated in some of the chapters in Whiten and Byrne (1997).

3. Evidence is mounting that great apes are cognitively more sophisticated than other primates. However, the Machiavellian intelligence hypothesis cannot help to explain the evolution of any cognitive attribute in great apes alone, because great apes simply do not form systematically larger groups than monkeys. To discover the origins of this extra sophistication we must relate their abilities to attributes of modern great apes not shown by monkeys, and especially to challenges in the physical environment that might have applied with particular severity to ancestral great apes rather than to monkeys.

Several theories have been proposed, all centering on the large body size of apes, relative to monkeys. Perhaps arboreal clambering, by large animals for whom a fall would be lethal, is best accomplished by advance planning, an engineering problem in which an ability to perceive the self as an object moving in space would be advantageous (Povinelli and Cant, 1995). Thus locomotion could directly have selected for imagination; however, as the orangutan's locomotion is unique, it is unclear how this proposal could be generalized and tested. Large size also indirectly puts a premium on efficient feeding, especially in direct competition with Old World monkeys, species able to digest a wider range of plant foods than great apes. Perhaps, then, the ability to process plant foods that are nutritious but difficult to extract from the matrix was selected in ancestral great apes; more narrowly, perhaps tool use was advantageous to the ancestral ape population for efficient extraction of certain embedded foods (Parker and Gibson, 1977; Gibson, 1990).

Relationships have yet to be demonstrated between higher intellectual comprehension and (a) the concept of embedding, presumably well understood by any food-storing bird or mammal, or (b) tool use, shown by numerous birds and mammals not noted for intelligence in other ways. Perhaps efficient food monopolization was enabled by a change to sleeping near food, involving the construction of arboreal platforms to give safety from predators (Fruth and Hohmann, 1996). The nests or beds constructed by all wild great apes for sleeping and resting are often elaborate structures showing skillful construction, although whether—as in birds—a strong genetic component is present is not known. Perhaps devising novel programs of manual actions for handling hard-to-eat foods enabled great apes to survive direct feeding competition from monkeys, whether or not these programs involve tools or extractive foraging. In extant mountain gorillas, food preparation programs are acquired rapidly during development and show complexity in structure and hierarchical organization (Byrne and Byrne, 1991, 1993; Byrne, in press). A sophisticated understanding of action would pay here, both for remembering organized programs of manual actions, and for understanding and learning from the actions of others (Byrne and Russon, 1998; see also Parker, 1996).

At present, all these theories are speculative and can be evaluated only by their relative plausibility, self-consistency, and partial match to the behavior of modern great apes. It would be premature to attempt to decide among them, but it may well be that physical, not social, challenges resulted in the evolution of mental representation capability among the great apes and ourselves.

7. Brains on Two Legs

1. The main ecological hypotheses for large brain size in primates are (a) that fruit-eating species (frugivores like spider monkeys) need larger brains than leaf-eating species (folivores like howler monkeys) because fruits are less predictable in time and space and therefore require more computing power to track; (b) that species with large home ranges (e.g., baboons) require bigger brains to manage the larger mental maps needed in order to navigate the home-range area than do species with small ranges (e.g., marmosets); and (c) that species with technically more complex diets ("extractive foragers" such as chimpanzees or capuchins) need larger brains to solve these technical problems than do species with less complex diets (nonextractive foragers like leaf monkeys). Extractive foraging is defined as those foraging strategies that involve having to remove food from some kind of matrix into which it is naturally embedded: examples include chimpanzees when they extract termites from termite mounds carefully in order to catch any at all, capuchin monkeys smashing open palm nuts in order to extract the meat inside, and humans having to hunt animals that are disinclined to sit around waiting to be caught.

2. The measure of relative neocortex size used here is the ratio of neocortex volume to volume of the rest of the brain. This ratio removes the confounding effects of changes in absolute brain (or even body) size: bigger-brained species would be expected to have bigger neocortices simply by virtue of their larger size. The substantive question is why some species have neocortices bigger than we might expect even for their large brain sizes.

3. A couple of logical points need to be clarified. First, it is important to appreciate that, evolutionarily speaking, it has been the need to evolve larger groups that has driven brain-size evolution in primates. The assumption here is that neocortex size limits group size because the size-dependent processing capacity of the neocortex constrains the kinds of complex social strategizing needed to keep a large group together as a coherent social unit. We cannot test an evolutionary hypothesis of this kind directly. Instead, what we can do is test its logical consequence—namely, the fact that the size of the neocortex imposes a constraint on group size in living species.

A second problem we might identify is the fact that plots of the kind shown in Figure 7.1 are, strictly speaking, correlations between two variables, and correlations cannot be used to infer causation. In other words, the causal relationship between group size and neocortex size might run in either direction: group size might determine brain size, or brain size might determine group size in living species. The first alternative is biologically implausible: brain size is much less labile developmentally than group size and it is difficult to imagine how some future ideal group size could conceivably determine the size of a species' brain at birth.

4. Two cautions should be noted. First, although there is a general relationship between group size and neocortex size in primates (and this relationship has been shown to extend to carnivores, cetaceans, bats, and some insectivores), there appear to be quite distinct grades within the order Primates. Prosimians are able to support larger groups for a given neocortex volume than monkeys can, and monkeys in turn can support larger groups than apes.

One reason why gorillas have large brains yet seem so asocial (gorilla group sizes typically average seven to ten individuals) is thus that they lie on a hominoid subgrade of the general relationship. Gorillas have a large brain but a small neocortex: their typical group sizes correspond closely to their neocortex sizes, but are much smaller than would be predicted by total brain sizes. Gorillas have a large cerebellum compared to other primates, probably because of the formidable coordination tasks associated with managing a very large body mass in trees.

The second point concerns the orangutan. This species seems at first sight to be anomalous, in that it has a large brain but is usually described as semisolitary. However, we need to remember, first, that a single counterexample cannot refute a statistical relationship, and second (and more important), that doubts have to be raised about both the neocortex volume and the characteristic group size of the orangutan. Unfortunately, no one has yet provided data on orangutan neocortex size; the lesson afforded by the gorilla should warn us not to make hasty assumptions on the basis of total brain size alone, especially since orangutans are more arboreal than Western lowland gorillas. In addition, we ought to question the conventional assumption that orangutans are solitary: the field data indicate that they are in fact more social than this conclusion implies (not only do some populations exhibit modest group sizes, but the members of a local resident population are obviously not strangers to one another). More interestingly perhaps, studies of two populations of group-living captive orangutans by Nilofer Ghaffar and Isabelle Lardeux-Gilloux reveal that they have social competencies and personality profiles more characteristic of the very social chimpanzees than the genuinely less social gorillas. The apparent asociality of the orangutan may thus be a consequence of the ecological conditions in its current retreat habitat on the margins of its former range.

5. For a summary, see Dunbar (1988). See also Hill and Dunbar (1998), Hill and Lee (1998), and the special 1998 issue of *Behaviour*, vol. 135(4).

8. From Primate Communication to Human Language

1. For more on generative grammar and innate language acquisition devices see Chomsky (1957) and Pinker (1994). Arguments for anatomic uniqueness are

presented by Lieberman (1975), Corballis (1991), and Hopkins et al. (1998); symbolic communication is stressed by Burling (1986) and by Noble and Davidson (1994). Bickerton (1990, 1995) emphasizes the role of self-awareness, and Dunbar (1998b) discusses the role of social complexity in language evolution. The design features of language are presented by Hockett (1963).

2. For birdsong as a parallel to language acquisition see Marler (1970), and for details on marmosets and tamarins see Snowdon, Elowson, and Roush (1997). For symbol-using great apes see Savage-Rumbaugh et al. (1993), Savage-Rumbaugh and Lewin (1994), and Matsuzawa (1996). For use of symbolic communication by dolphins see Herman, Pack, and Morrel-Samuels (1993), and for parrots see Pepperberg (1997). Research on symbolic communication in vervets is presented by Cheney and Seyfarth (1990a); syntax in capuchin monkeys by Robinson (1984), in cotton-top tamarins by Cleveland and Snowdon (1982), and McConnell and Snowdon (1986), and in pygmy marmosets by Addington (1998). Chickadee syntax is presented in Hailman, Ficken, and Ficken (1985). Evidence for birds learning song of other species is presented by Baptista and Gaunt (1997), for macaques understanding the calls of other species by Seyfarth and Cheney (1997), for vervets responding to starling alarm calls by Hauser (1988), and for lemurs responding to sifaka alarms by Oba and Masataka (1996). Herman et al. (1993) and Matsuzawa (1996) describe duality of patterning in dolphins and chimpanzees respectively. Lieberman (1975) describes chimpanzee vocal anatomy. Maddieson (1984) describes variability in human phonemes. Categorical perception of speech in humans is presented by Liberman et al. (1967), in macaques by Morse and Snowdon (1975) and Waters and Wilson (1976), in chinchillas by Kuhl and Miller (1975), and in Japanese quail by Kluender, Diehl, and Killeen (1987) and Kluender (1994). For categorical perception of animals' own sounds, see Nelson and Marler (1989) for swamp sparrows; Ehret (1987) for mice; May, Moody, and Stebbins (1989) and Owren, Seyfarth, and Hopp (1992) for macaques; Masataka (1983) for titi monkeys; and Snowdon and Pola (1978) for pygmy marmosets. Snowdon (1987) provides a naturalistic view of categorical perception. Motivational-structural rules are described in Owings and Morton (1998), and how humans apply these structures to communication with animals is described by McConnell (1990) and with infants by Fernald (1992). Van Hooff (1972) discusses the evolution of smiling and laughing from monkeys to humans; Parr, Dove, and Hopkins (1998) and Parr, Hopkins, and de Waal (1999) describe how chimpanzees perceive facial expressions; and Kanazawa (1998) describes how macaques perceive emotions in human faces.

3. The language instinct is argued by Pinker (1994). Jakobson and Halle (1956) and Greenberg (1963) discuss language universals, while diversity in language sounds is presented by Maddieson (1984) and Kluender (1994). Bates and Marchman (1988) give a critique of language universals in development. The study of Chinese and Koreans learning English as a second language is by Johnson and Newport (1989), and the multilingual cultures in the Amazon are described by Sorensen (1967). Saffran, Aslin, and Newport (1996) show how in-

fants can extract "words" from strings of nonsense syllables. Huttenlocher et al. (1991) demonstrate the importance of input to language acquisition, and Locke and Snow (1997) review studies of language development in abused and isolated infants. The study of language development in twins is by Jouanjean-l'Antoene (1997).

4. Birdsong development is described by Marler and Peters (1982) and Baptista and Gaunt (1997), in gibbons by Brockelman and Schilling (1984) and Tenaza (1985). Snowdon (1993) provides an example of using primate long calls in taxonomy. Studies of the effects of isolation rearing or deafening on monkey vocal development are found in Talmage-Riggs et al. (1972), Newman and Symmes (1982), and Herzog and Hopf (1984). Seyfarth and Cheney (1997) provide a comprehensive review of development in nonhuman primates. Examples of birds altering song or call structure in response to changes in social companions are found in Mundinger (1970); Nowicki (1989); Farabaugh, Linzenbold, and Dooling (1994); and Hausberger et al. (1995). Pygmy marmoset trills and their use by adults are described by Pola and Snowdon (1975), Snowdon and Hodun (1981), and Snowdon and Cleveland (1984). Ontogeny of trills is presented in Elowson, Sweet, and Snowdon (1992) and Elowson and Snowdon (1994). Trill convergence with mated pairs is discussed in Snowdon and Elowson (1999). Call convergence or divergence in response to changes in social environments is reported by Randall (1995) for kangaroo rats, Boughman (1997, 1998) for spear-nosed bats, Tyack and Sayigh (1997) for dolphins, Stanger (1993) for mouse lemurs, and Giles and Smith (1979) for humans. Babbling and its function for human infants is presented in de Boysson-Bardies et al. (1992) and Locke (1993). Petitto and Martentette (1991) describe manual babbling in deaf children. Elowson, Snowdon, and Lazaro-Perea (1998a and b) describe pygmy marmoset babbling, and Omedes (1985) describes a similar phenomenon in silvery marmosets. Elowson, Tannenbaum, and Snowdon (1991) describe the use of food calls in adult cotton-top tamarins; Roush and Snowdon (1994) describe the development of these calls; and Roush and Snowdon (in press) show how food-call structure and usage changes when animals are paired.

5. Bickerton (1990) describes the single-gene mutation hypothesis. Hemelrijk (1996) provides models of self-organizing or bottom-up principles of reciprocal altruism. Self-produced song in zebra finches is presented in Volman and Khanna (1995), the emergence of gestural sequences in deaf children by Goldin-Meadow (1997), and in the Nicaraguan deaf community by Kegl, Senghas, and Coppola (1996). Matsuzawa (1996) describes the hierarchical structure of tool use in chimpanzees, and Toth and Schick (1993) describe the syntax of hominid tool production. Cheney and Seyfarth (1990a) and King (1994) discuss social learning of communication and other skills in monkeys. Mitani (1996) reviews current work on vocal complexity in apes, and Leavens and Hopkins (1999) review work on pointing in monkeys and apes. Armstrong, Stokoe, and Wilcox (1995) describe gesture and the evolution of human sign languages, and Corballis (1999) presents a modern version of the gestural origins of language. McCune et al. (1996) describe grunting as an indicative act in human infants.

Dunbar (1997) presents the gossip hypothesis of language origins, and Stanford (1999) describes hunting in apes and humans.

9. The Nature of Culture

1. For an accessible account of the evolution of human technology, see Schick and Toth (1993). The widest-ranging survey of culture in animals is that of Bonner (1980), who uses the concept so generally that it is applied to phyla from slime molds to apes. For material culture, Beck's (1980) encyclopedic monograph on tool use and tool-making remains the standard work. The burgeoning field of orca or killer whale *(Orcinus orca)* culturology is exemplified by Barrett-Leonard et al. (1996); Diamond's (1987) ethnography of highland New Guinean bowerbirds (Ptilonorhynchidae) remains the most detailed; the interlocking field and laboratory work of Terkel (1996) and his students on black rats *(Rattus rattus)* colonizing Israeli agro-forestry plantations is summarized admirably. Snowdon and Hausberger (1997) present a collection of articles on social learning of vocal communication in birds and mammals.

For a specific account of the most complex and well-studied cultural phenomenon apart from humans, that is, song-learning by passerine birds, see West and King (1996) on cowbirds *(Molothrus ater)*. These brood parasites lay their eggs in the nests of other species, so the youngsters must learn to sing the species-specific and population-specific songs of their genetic parents (whom they never meet) and ignore the songs of their foster parents. How they do so is complex. It turns out that male cowbirds learn their local dialects from nonsinging females, who shape the males by differential behavioral reinforcement!

2. Tylor (1871) is usually credited as being the founder of American anthropology; his oft-quoted definition of culture appears in most textbooks. Kroeber's (1928) musing on nonhuman culture came in response to his eye-opening reading of Köhler's (1927) masterwork (see below). Harris (1979) has considered over decades the question of human versus nonhuman culture and has come to varying conclusions. The skepticism of laboratory psychologists toward social learning in nonhuman species is exemplified by Galef (1992), who mainly studies food preferences in rats. The laboratory experiments on children and great apes of Tomasello and Call (1997) and much more are synthesized in their 1998 book. Matsuzawa's (1994) field experiments at Bossou on West African chimpanzees have been done in collaboration with Sugiyama (1997), a pioneering field-worker in primatology. De Waal's (1982) best-selling book on the political lives of the captive chimpanzee colony at Arnhem Zoo was revised in 1998. Anderson (1996) is one of the few primatologists who has worked with more than one species of capuchin *(Cebus* spp.) monkey; no one seems to have done prolonged research on the genus in both captivity and nature. McGrew and Marchant (1997) systematically discuss anecdotes, idiosyncrasies, and the like in comparing the tool use and laterality of hand function in capuchin monkeys and chimpanzees *(Pan* spp.).

3. The most recent review of culture in nonhuman primates is by McGrew (1998), but Nishida (1987) supplies a fuller account of earlier studies. Köhler's 1927 book (published earlier in German) has inspired generations of students of chimpanzees, and his problem-solving tasks, such as box stacking and raking in out-of-reach objects, are standards. Itani and Nishimura (1973) were the first to summarize in English studies of culture in Japanese macaques *(Macaca fuscata)*; Itani is a student of Imanishi. Suzuki et al. (1990) describe studies of the wilderness-living monkeys of Yakushima. For decades Altmann and Altmann (1970) have studied the yellow (savanna) baboons *(Papio cynocephalus)* of Amboseli, Kenya. Kummer (1995) has studied desert baboons *(Papio hamadryas)* in zoos and nature for even longer. Long-running studies of capuchin monkeys (mostly *Cebus apella*) in captivity have been centered at Strasbourg (Anderson, 1996), Rome (Visalberghi 1997), and Poolesville, Maryland (Westergaard 1998). Fedigan's (1990) studies of wild *Cebus capucinus* at Santa Rosa, Costa Rica, exemplify long-term fieldwork on this genus. The most habituated wild bonobos *(Pan paniscus)* are those at Wamba, Democratic Republic of Congo (Kano, 1992); the most comprehensive account of the species is a book by de Waal (1997a). The longest and best-known study of wild orangutans *(Pongo pygmaeus)* is that of Galdikas (1995) at Tanjung Puting, Kalimantan (Borneo), Indonesia. Fox et al. (1998) present the results on elementary technology of wild orangutans in Sumatra. For Russon's study of rehabilitant orangutans imitating humans, see Russon and Galdikas (1995). Byrne (1996a) examines the enigmatic psyche of the gorilla *(Gorilla gorilla)*. The longest-running field study of lowland gorillas is that of Tutin and Fernandez (1992) at Lopé, in Gabon. For comprehensive syntheses of the socioecology of great apes, see the chapters by van Schaik (orangutan), Watts (gorilla), and White (bonobo) in McGrew et al. (1996).

4. For the ethnography of chimpanzees, the magnum opus for Gombe is Goodall (1986), for Mahale is Nishida (1990), for Taï is Boesch and Boesch (2000). For Bossou there is no single book, but Matsuzawa (1996) and Sugiyama (1997) review the progress. The most comprehensive evaluative synthesis of material culture in *Pan troglodytes* is McGrew (1992). An up-to-date and exhaustive cross-cultural catalog of chimpanzees compares six key field sites (Whiten et al., 1999); this includes a useful reference map of the distribution of customs. For an extensive analysis of nest building in great apes, see Fruth and Hohmann (1996). Wrangham et al. (1994) discuss behavioral flexibility. Dunbar (1996) speculates on the evolutionary significance to humans of social grooming in primates. For details of the grooming handclasp, see McGrew (1992); for social scratch, see Nakamura et al. (1999). Kortlandt (1986) focused on the Nimba region of West Africa and, based on his extensive field experience, speculated on the significance of chimpanzees using stone tools. Boesch et al. (1994) established the Sassandra River as the eastern boundary of cultural diffusion of nutcracking by chimpanzees. McGrew et al. (1997) performed ecological surveys in Lopé Reserve, Gabon, of nut-bearing trees, and of the wooden and stone raw materials for hammers and anvils. Few archaeologists have worked in the field side by side with primatologists; Joulian (1996) did so with Boesch at Taï.

5. The most useful recent compendium on ape versus human culture is Boesch and Tomasello's 1998 article; see also comments on this article in the same journal by Byrne, Galef, Ingold, McGrew, Paterson, and Whiten. The best single treatment of ape intelligence is that of Byrne (1995). Whiten and Ham (1992) compiled a classificatory review of imitation, and Whiten (1998) exemplifies careful empirical treatment of the topic in apes. Byrne and Russon (1998) and commentators air the issues regarding imitation in primates. For teaching in animals, see Caro and Hauser (1992) and Maestripieri (1995). For studies of American Sign Language in chimpanzees, see Fouts (1994). For field studies comparing communication of wild apes, both across species and across populations, see Mitani (1996) for the long calls of African apes and Arcadi et al. (1998) for vocalizations and drumming by chimpanzees. Data on the progressive ratchet effect in the macaques of Koshima, on both wheat sluicing and sweet-potato washing, are in Kawai et al. (1992) and Watanabe (1994). Trivers (1985) provided the evolutionary biological framework in which to consider culture, given that all social interactions amount to altruism, competition, cooperation, or spite. No one has discussed the costs of imitation and teaching, especially as applied to culture, but see Russon (1997). Hewlett and Cavalli-Sforza (1986) systematically analyzed social transmission in the daily lives of African foragers, the Aka of the Congo basin; more such comparative studies are needed.

Bibliography

Adams, D., et al. 1986. The Seville Statement on Violence. *American Psychologist,* 45: 1167–68.

Addington, R. L. 1998. Social foraging in captive pygmy marmosets *(Cebuella pygmaea).* Effects of food characteristics and social context on feeding competition and vocal behavior. Unpublished doctoral dissertation, University of Wisconsin, Madison.

Aiello, L. C., and Dunbar, R. I. M. 1993. Neocortex size, group size and the evolution of language. *Current Anthropology,* 34: 184–193.

Aiello, L. C., and Wheeler, P. 1995. The expensive tissue hypothesis. *Current Anthropology,* 36: 199–211.

Altmann, J., Alberts, S. C., et al. 1996. Behavior predicts genetic structure in a wild primate group. *Proceedings of the National Academy of Sciences USA,* 93: 5797–5801.

Altmann, S. A., and Altmann, J. 1970. *Baboon Ecology.* Basel: S. Karger.

Anderson, J. R. 1984. Monkeys with mirrors: Some questions for primate psychology. *International Journal of Primatology,* 5: 81–98.

——— 1996. Chimpanzees and capuchin monkeys: Comparative cognition. In *Reaching into Thought: The Minds of the Great Apes* (A. E. Russon, K. A. Bard, and S. T. Parker, eds.), pp. 23–56. Cambridge: Cambridge University Press.

Arcadi, A. C., Robert, D., and Boesch, C. 1998. Buttress drumming by wild chimpanzees: Temporal patterning, phrase integration into loud calls, and preliminary evidence for individual distinctiveness. *Primates,* 39: 505–518.

Ardrey, R. 1961. *African Genesis.* New York: Atheneum.

Armstrong, D. F., Stokoe, W. C., and Wilcox, S. E. 1995. *Gesture and the Nature of Language.* Cambridge: Cambridge University Press.

Aureli, F., and de Waal, F. B. M. (2000). *Natural Conflict Resolution.* Berkeley: University of California Press.

Badrian, A., and Badrian, N. 1984. Social organization of *Pan paniscus* in Lomako Forest, Zaire. In *The Pygmy Chimpanzee* (R. L. Susman, ed.), pp. 325–346. New York: Plenum.

Bailey, R. C. 1985. The socioecology of Efe pygmy men in the Ituri Forest, Zaïre. Unpublished doctoral dissertation, Harvard University, Cambridge, Mass.

Bailey, R. C., and Peacock, N. 1988. Efe pygmies of Northeast Zaïre: Subsistence strategies in the Ituri Forest. In *Coping with Uncertainty in Food Supply* (I. de Garine and G. Harrison, eds.), pp. 88–117. Oxford: Oxford University Press.

Baptista, L. F., and Gaunt, S. L. L. 1997. Social interaction and vocal development in birds. In *Social Influences on Vocal Development* (C. T. Snowdon and M. Hausberger, eds.), pp. 23–40. Cambridge: Cambridge University Press.

Barrett-Leonard, L. G., Ford, J. K. B., and Heise, K. A. 1996. The mixed blessing of echolocation: Differences in sonar use by fish-eating and mammal-eating killer whales. *Animal Behaviour*, 51: 553–568.

Bartlett, T. Q., Sussman, R. W., and Cheverud, J. M. 1993. Infant killing in primates: A review of observed cases with specific reference to the sexual selection hypothesis. *American Anthropologist*, 95: 958–990.

Barton, R. A. 1996. Neocortex size and behavioural ecology in primates. *Proceedings of the Royal Society, London, B*, 263: 173–177.

Barton, R. A., and Dunbar, R. I. M. 1996. Evolution of the social brain. In *Machiavellian Intelligence, II: Extensions and Evaluations* (A. Whiten and R. Byrne, eds.), pp. 240–263. Cambridge: Cambridge University Press.

Bates, E., and Marchman, V. A. 1988. What is and is not universal in language acquisition. In *Language, Communication and the Brain* (F. Plum, ed.), pp. 19–38. New York: Raven Press.

Beck, B. B. 1980. *Animal Tool Behavior.* New York: Garland STPM Press.

Berkowitz, L. 1993. *Aggression: Its Causes, Consequences, and Control.* New York: McGraw-Hill.

Bickerton, D. 1990. *Language and Species.* Chicago: University of Chicago Press.

——— 1995. *Language and Human Behavior,* Seattle: University of Washington Press.

Blumenschine, R. J. 1987. Characteristics of an early hominid scavenging niche. *Current Anthropology*, 28: 383–407.

Boehm, C. 1992. Segmentary "warfare" and the management of conflict: Comparison of East African chimpanzees and patrilineal-patrilocal humans. In *Coalitions and Alliances in Humans and Other Animals* (A. H. Harcourt and F. B. M. de Waal, eds.), pp. 137–174. Oxford: Oxford University Press.

Boesch, C. 1991. The effects of leopard predation on grouping patterns in forest chimpanzees. *Behaviour*, 117: 220–242.

——— 1994. Cooperative hunting in wild chimpanzees. *Animal Behaviour*, 48: 653–667.

——— 1996. Social grouping in Taï chimpanzees. In *Great Ape Societies* (W. C. McGrew, L. F. Marchant and T. Nishida, eds.), pp. 101–113. Cambridge: Cambridge University Press.

Boesch, C., and Boesch, H. 1989. Hunting behavior of wild chimpanzees in the Taï National Park. *American Journal of Physical Anthropology*, 78: 547–573.

——— 1994. Hominization in the rainforest: The chimpanzee's piece of the puzzle. *Evolutionary Anthropology*, 3: 9–16.

——— 2000. *The Chimpanzees of Taï: Behavioural Ecology and Evolution.* Oxford: Oxford University Press.

Boesch, C., Marchesi, P., Marchesi, N., Fruth, B., and Joulian, F. 1994. Is nut cracking in wild chimpanzees a cultural behavior? *Journal of Human Evolution*, 26: 325–338.

Boesch, C., and Tomasello, M. 1998. Chimpanzee and human cultures. *Current Anthropology*, 39: 591–614.

Bonner, J. T. 1980. *The Evolution of Culture in Animals*. Princeton: Princeton University Press.

Boughman, J. W. 1997. Greater spear-nosed bats give group distinctive calls. *Behavioral Ecology and Sociobiology*, 40: 61–70.

—— 1998. Vocal learning by greater spear-nosed bats. *Proceedings of the Royal Society, London, B*, 265: 227–233.

Boysen, S. T. 1998. Attribution processes in chimpanzees. Heresy, hearsay or heuristic? In *17th Congress of the International Primatological Society*. University of Antananarivo, Madagascar, August 10–14.

Broadbent, D. E. 1961. *Behaviour*. London: Methuen.

Brockelman, W. Y., and Schilling, D. 1984. Inheritance of stereotyped gibbon calls. *Nature*, 312: 634–636.

Bunn, H. T., and Ezzo, J. A. 1993. Hunting and scavenging by Plio-Pleistocene hominids: Nutritional constraints, archaeological patterns, and behavioural implications. *Journal of Archaeological Science*, 20: 365–398.

Burling, R. 1986. The selective advantage of complex language. *Ethology and Sociobiology*, 7: 1–16.

Bygott, J. D. 1979. Agonistic behaviour and dominance among wild chimpanzees. In *The Great Apes* (D. A. Hamburg and E. R. McCown, eds.), pp. 405–427. Menlo Park, Calif.: Benjamin/Cummings.

Byrne, R. W. 1995. *The Thinking Ape: Evolutionary Origins of Intelligence*. Oxford: Oxford University Press.

—— 1996a. The misunderstood ape: Cognitive skills of the gorilla. In *Reaching into Thought: The Minds of the Great Apes* (A. E. Russon, K. A. Bard, and S. T. Parker, eds.), pp. 111–130. Cambridge: Cambridge University Press.

—— 1996b. Relating brain size to intelligence. In *Modelling the Early Human Mind* (P. A. Mellars and K. R. Gibson, eds.), pp. 49–56. Cambridge: McDonald Institute for Archaeological Research.

—— 1997. The technical intelligence hypothesis: An additional evolutionary stimulus to intelligence? In *Machiavellian Intelligence, II: Extensions and Evaluations* (A. Whiten and R. W. Byrne, eds.), pp. 289–311. Cambridge: Cambridge University Press.

—— In press. Cognition in great ape ecology. Skill-learning ability opens up foraging opportunities. *Symposia of the Zoological Society of London*.

Byrne, R. W., and Byrne, J. M. E. 1991. Hand preferences in the skilled gathering tasks of mountain gorillas *(Gorilla g. beringei)*. *Cortex*, 27: 521–546.

—— 1993. Complex leaf-gathering skills of mountain gorillas *(Gorilla g. beringei)*: Variability and standardization. *American Journal of Primatology*, 31: 241–261.

Byrne, R. W., and Russon, A. E. 1998. Learning by imitation: A hierarchical approach. *Behavioral and Brain Sciences*, 21: 667–721.

Byrne, R. W., and Whiten, A. 1985. Tactical deception of familiar individuals in baboons *(Papio ursinus)*. *Animal Behaviour*, 33: 669–673.

—— 1988. *Machiavellian Intelligence: Social Expertise and the Evolution of Intellect in Monkeys, Apes and Humans*. Oxford: Clarendon Press.

—— 1990. Tactical deception in primates: The 1990 data-base. *Primate Report*, 27: 1–101.

—— 1992. Cognitive evolution in primates: Evidence from tactical deception. *Man*, 27: 609–627.

Call, J., and Tomasello, M. 1998. Distinguishing intentional from accidental actions in orangutans *(Pongo pygmaeus)*, chimpanzees *(Pan troglodytes)*, and human children *(Homo sapiens)*. *Journal of Comparative Psychology*, 112: 192–206.

Caro, T. M., and Hauser, M. D. 1992. Is there teaching in nonhuman animals? *Quarterly Review of Biology*, 67: 151–174.

Cartmill, M. 1993. *A View to a Death in the Morning: Hunting and Nature through History*. Cambridge, Mass.: Harvard University Press.

Chance, M. R. A., and Mead, A. P. 1953. Social behaviour and primate evolution. *Symposia of the Society of Experimental Biology*, 7: 395–439.

Chaplin, G., Jablonski, N. G., and Cable, N. T. 1994. Physiology, thermoregulation and bipedalism. *Journal of Human Evolution*, 27: 497–510.

Cheney, D. L., and Seyfarth, R. M. 1983. Non-random dispersal in free-ranging vervet monkeys: Social and genetic consequences. *American Naturalist*, 122: 392–412.

—— 1986. The recognition of social alliances among vervet monkeys. *Animal Behaviour*, 34: 1722–31.

—— 1990a. *How Monkeys See the World*. Chicago: University of Chicago Press.

—— 1990b. Attending to behaviour versus attending to knowledge: Examining monkeys' attribution of mental states. *Animal Behaviour*, 40: 742–753.

Chomsky, N. 1957. *Syntactic Structures*. The Hague: Mouton.

—— 1959. Review of Skinner, 1957. *Language*, 35: 26–58.

Clarke, R. J., and Tobias, P. V. 1995. Sterkfontein member 2 foot bones of the oldest South African hominid. *Science*, 269: 521–524.

Cleveland, J., and Snowdon, C. T. 1982. The complex vocal repertoire of the adult cotton-top tamarin *(Saguinus oedipus)*. *Zeitschrift für Tierpsychologie*, 58: 231–270.

Clutton-Brock, T. H. 1989. Female transfer and inbreeding avoidance in social mammals. *Nature*, 337: 70–72.

Coolidge, H. J. 1933. *Pan paniscus*: Pygmy chimpanzee from south of the Congo River. *American Journal of Physical Anthropology*, 18: 1–57.

Corballis, M. C. 1991. *The Lopsided Ape: Evolution of the Generative Mind*. New York: Oxford University Press.

—— 1999. The gestural origins of language. *American Scientist*, 87: 138–145.

Corballis, M. C., and Lea, S. E. G. 1999. *The Descent of Mind: Psychological Perspectives on Hominid Evolution.* Oxford: Oxford University Press.

Cords, M. 1997. Friendships, alliances, reciprocity and repair. In *Machiavellian Intelligence, II: Extensions and Evaluations* (A. Whiten and R. W. Byrne, eds.), pp. 24–49. Cambridge: Cambridge University Press.

Dart, R. A. 1925. *Australopithecus africanus:* The man-ape of South Africa. *Nature,* 115: 195–199.

Darwin, C. 1871. *The Descent of Man and Selection in Relation to Sex.* London: J. Murray.

Dasser, V. 1988. Mapping social concepts in monkeys. In *Machiavellian Intelligence: Social Expertise and the Evolution of Intellect in Monkeys, Apes and Humans* (R. W. Byrne and A. Whiten, eds.), pp. 85–93. Oxford: Clarendon Press.

de Boysson-Bardies, B., Vihman, M. M., Roug-Hellichius, L., Durand, C., Landberg, I., and Arao, F. 1992. Material evidence of infant selection from the target language: A cross-linguistic phonetic study. In *Phonological Development: Models, Research, Implications* (C. Ferguson, L. Menn, and C. Stoel-Gammon, eds.), pp. 369–391. Timonium, Md.: York Press.

Diamond, J. 1987. Bower building and decoration by the bowerbird, *Amblyornis inornatus. Ethology,* 74: 177–204.

———— 1992. *The Third Chimpanzee: The Evolution and Future of the Human Animal.* New York: Harper Collins.

Dickmann, M. 1981. Paternal confidence and dowry competition: A biocultural analysis of purdah. In *Natural Selection and Social Behavior* (R. D. Alexander and D. W. Tinkle, eds.), pp. 417–438. New York: Chiron Press.

Di Fiore, A., and Rendall, D. 1994. Evolution of social organization: A reappraisal for primates by using phylogenetic methods. *Proceedings of the National Academy of Sciences USA,* 91: 9941–45.

Doran, D. 1997. Influence of seasonality on activity patterns, feeding behavior, ranging, and grouping patterns in Taï chimpanzees. *International Journal of Primatology,* 18: 183–206.

Dunbar, R. I. M. 1988. *Primate Social Systems.* London: Croom Helm.

———— 1991. Functional significance of social grooming in primates. *Folia Primatology,* 57: 121–131.

———— 1992a. Time: A hidden constraint on the behavioural ecology of baboons. *Behavioural Ecology and Sociobiology,* 31: 35–49.

———— 1992b. Neocortex size as a constraint on group size in primates. *Journal of Human Evolution,* 20: 469–493.

———— 1992c. Behavioural ecology of the extinct papionines. *Journal of Human Evolution,* 22: 407–421.

———— 1993. Socioecology of the extinct theropiths: A modelling approach. In *Theropithecus: The Rise and Fall of a Primate Genus* (G. Jablonski, ed.), pp. 465–486. Cambridge: Cambridge University Press.

———— 1996. Determinants of group size in primates: A general model. *Proceedings of the British Academy,* 88: 33–57.

———— 1997. *Grooming, Gossip and the Evolution of Language.* London: Faber and Faber.

———— 1998a. The social brain hypothesis. *Evolutionary Anthropology,* 6: 178–190.

———— 1998b. Theory of mind and the evolution of language. In *Approaches to the Evolution of Language* (J. R. Hurford, M. Studdert-Kennedy, and C. Knight, eds.), pp. 92–110. Cambridge: Cambridge University Press.

Dunbar, R. I. M., Duncan, N. D. C., and Marriott, A. 1997. Human conversational behaviour. *Human Nature,* 8: 231–246.

Dunbar, R. I. M., Duncan, N. D. C., and Nettle, D. 1995. Size and structure of freely forming conversational groups. *Human Nature,* 6: 67–78.

Ehret, G. 1987. Categorical perception of sound signals: Facts and hypotheses from animal studies. In *Categorical Perception* (S. Harnad, ed.), pp. 301–331. New York: Cambridge University Press.

Elowson, A. M., and Snowdon, C. T. 1994. Pygmy marmosets, *Cebuella pygmaea,* modify vocal structure in response to changed social environment. *Animal Behaviour,* 47: 1267–77.

Elowson, A. M., Snowdon, C. T., and Lazaro-Perea, C. 1998a. "Babbling" and social context in infant monkeys: Parallel to human infants. *Trends in Cognitive Science,* 2: 35–43.

———— 1998b. Infant "babbling" in a nonhuman primate: Complex sequences of vocal behavior. *Behaviour,* 135: 643–664.

Elowson, A. M., Sweet, C. S., and Snowdon, C. T. 1992. Ontogeny of trill and J-call vocalizations in the pygmy marmoset, *Cebuella pygmaea. Animal Behaviour,* 43: 703–715.

Elowson, A. M., Tannenbaum, P. L., and Snowdon, C. T. 1991. Food associated calls correlate with food preferences in cotton-top tamarins. *Animal Behaviour,* 42: 931–937.

Farabaugh, S. M., Linzenbold, A., and Dooling, R. J. 1994. Vocal plasticity in budgerigars *(Melopsittacus undulatus):* Evidence for social factors in the learning of contact calls. *Journal of Comparative Psychology,* 108: 81–92.

Fedigan, L. M. 1990. Vertebrate predation in *Cebus capucinus:* Meat eating in a neotropical monkey. *Folia Primatologica,* 54: 196–205.

Fernald, A. 1992. Human maternal vocalizations to infants as biologically relevant signals: An evolutionary perspective. In *The Adapted Mind* (J. Barkow, L. Cosmides, and J. Tooby, eds.), pp. 391–428. New York: Oxford University Press.

Fisher, H. 1983. *The Sex Contract: The Evolution of Human Behavior.* New York: Quill.

Foley, R. A. 1995. *Humans before Humanity.* Oxford: Blackwell.

———— 1989. The evolution of hominid social behaviour. In *Comparative Socioecology* (V. Standen and R. A. Foley, eds.), pp. 473–494. Oxford: Blackwell Scientific.

Foley, R. A., and Lee, P. C. 1989. Finite social space, evolutionary pathways, and reconstructing hominid behavior. *Science,* 243: 901–906.

Fouts, D. H. 1994. The use of remote video recordings to study the use of American Sign Language by chimpanzees when no humans are present. In *The Ethological Roots of Culture* (R. A. Gardner, B. T. Gardner, B. Chiarelli, and F. X. Plooij, eds.), pp. 271–284. Dordrecht: Kluwer.

Fox, E. A., Sitompul, A. F., and van Schaik, C. P. 1998. Intelligent tool use in wild Sumatran orangutans. In *The Mentality of Gorillas and Orangutans* (S. T. Parker, H. L. Miles, and R. W. Mitchell, eds.). Cambridge: Cambridge University Press.

Fox, R. 1988. The Seville Declaration: Anthropology's auto-da-fé. *Academic Questions*, 1: 35–47.

Fruth, B., and Hohmann, G. 1996. Nest building in the great apes: The great leap forward? In *Great Ape Societies* (W. C. McGrew, L. F. Marchant, and T. Nishida, eds.), pp. 225–240. Cambridge: Cambridge University Press.

Furuichi, T. 1989. Social interactions and the life history of female *Pan paniscus* in Wamba, Zaire. *International Journal of Primatology*, 10: 173–197.

——— 1992. The prolonged estrus of females and factors influencing mating in a wild group of bonobos *(Pan paniscus)* in Wamba, Zaire. In *Topics in Primatology*, Vol. 2, *Behavior, Ecology, and Conservation* (N. Itoigawa, Y. Sugiyama, G. P. Sackett, and R. K. R. Thompson, eds.), pp. 179–190. Tokyo: University of Tokyo Press.

——— 1997. Agonistic interactions and matrifocal dominance rank of wild bonobos *(Pan paniscus)* at Wamba. *International Journal of Primatology*, 18: 855–875.

Furuichi, T., Idani, G., Ihobe, H., Kuroda, S., Kitamura, K., Mori, A., Enomoto, T., Okayasu, N., Hashimoto, C., and Kano, T. 1998. Population dynamics of wild bonobos *(Pan paniscus)* at Wamba. *International Journal of Primatology*, 19: 1029–43.

Gagneaux, P., Boesch, C., et al. 1999. Female reproductive strategies, paternity and community structure in wild West African chimpanzees. *Animal Behaviour*, 57: 19–32.

Galdikas, B. M. F. 1995. *Reflections of Eden*. London: Victor Gollancz.

Galef, B. G. 1992. The question of animal culture. *Human Nature*, 3: 157–178.

Gallup, G. G., Jr. 1970. Chimpanzees: Self-recognition. *Science*, 167: 86–87.

——— 1982. Self-awareness and the emergence of mind in primates. *American Journal of Primatology*, 2: 237–248.

Garber, P. A. 1994. Phylogenetic approach to the study of tamarin and marmoset social systems. *American Journal of Primatology*, 34: 199–219.

Ghiglieri, M. P. 1999. *The Dark Side of Man: Tracing the Origins of Violence*. New York: Perseus.

Gibson, K. R. 1990. New perspectives on instincts and intelligence: Brain size and the emergence of hierarchical mental construction skills. In *'Language' and Intelligence in Monkeys and Apes* (S. T. Parker and K. R. Gibson, eds.), pp. 97–128. Cambridge: Cambridge University Press.

Giles, H., and Smith, P. 1979. Accommodation theory: Optimal levels of convergence. In *Language and Social Psychology* (H. Giles and R. St. Clair, eds.), pp. 45–65. Oxford: Basil Blackwell.

Goldberg, T. L., and Wrangham, R. W. 1997. Genetic correlates of social behavior in wild chimpanzees: Evidence from mitochondrial DNA. *Animal Behaviour,* 54: 559–570.

Goldin-Meadow, S. 1997. The resilience of language in humans. In *Social Influences on Vocal Development* (C. T. Snowdon and M. Hausberger, eds.), pp. 293–311. Cambridge: Cambridge University Press.

Goodall, J. 1968. The behaviour of free-living chimpanzees of the Gombe Stream Reserve. *Animal Behaviour Monographs,* 1: 161–311.

——— 1971. *In the Shadow of Man.* Boston: Houghton Mifflin.

——— 1977. Infant killing and cannibalism in free-living chimpanzees. *Folia Primatologica,* 28: 259–282.

——— 1986. *The Chimpanzees of Gombe: Patterns of Behavior.* Cambridge, Mass.: Belknap Press of Harvard University Press.

——— 1990. *Through a Window.* Boston: Houghton Mifflin.

Goodall, J., Bandora, A., et al. 1979. Intercommunity interactions in the chimpanzee population of the Gombe National Park. In *The Great Apes* (D. A. Hamburg and E. R. McCown, eds.). Menlo Park, Calif.: Benjamin/Cummings.

Greenberg, J. H. 1963. *Universals of Language.* Cambridge, Mass.: MIT Press.

Groves, C. P. 1986. Systematics of the great apes. In *Comparative Primate Biology,* Vol. 1, *Systematics, Evolution and Anatomy* (D. R. Swindler, and J. Erwin, eds.), pp. 187–217. New York: Alan R. Liss.

Hailman, J. P., Ficken, M. S., and Ficken, R. W. 1985. The "chick-a-dee" call of *Parus atricapillus:* A recombinant system of animal communication compared with written English. *Semiotica,* 56: 191–224.

Hamilton, W. D. 1964. The genetical evolution of social behavior. *Journal of Theoretical Biology,* 7: 1–51.

Harcourt, A. 1992. Coalitions and alliances: Are primates more complex than non-primates? In *Coalitions and Alliances in Humans and Other Animals* (A. H. Harcourt and F. B. M. de Waal, eds.), pp. 445–471. Oxford: Oxford University Press.

Harcourt, A. H., and de Waal, F. B. M. (eds.) 1992. *Coalitions and Alliances in Humans and Other Animals.* Oxford: Oxford University Press.

Harcourt, A. H., Harvey, P. H., Larson, S. G., and Short, R. V. 1981. Testis weight, body weight, and breeding system in primates. *Nature,* 293: 55–57.

Harris, M. 1979. *Cultural Materialism.* New York: Vintage.

Hashimoto, C., and Furuichi, T. 1994. Social role and development of noncopulatory sexual behavior of wild bonobos. In *Chimpanzee Cultures* (R. W. Wrangham, W. C. McGrew, F. B. M. de Waal, and P. Heltne, eds.), pp. 155–168. Cambridge, Mass.: Harvard University Press.

Hausberger, M., Richard-Yris, M. A., Henry, L., Lepage, L., and Schmidt, I. 1995. Song sharing reflects the social organization in a captive group of European starlings *(Sturnus vulgaris). Journal of Comparative Psychology,* 109: 222–241.

Hauser, M. D. 1988. How vervet monkeys learn to recognize starling alarm calls: The role of experience. *Behaviour,* 105: 187–201.

Hawkes, K. 1991. Showing off: Tests of an hypothesis about men's foraging goals. *Ethology and Sociobiology,* 12: 29–54.

———— 1999. Hadza hunting and the evolution of nuclear families. *Current Anthropology* (submitted).

Hawkes, K., O'Connell, J. F., and Blurton-Jones, N. G. 1991. Hunting income patterns among the Hadza: Big game, common goods, foraging goals and the evolution of the human diet. *Philosophical Transactions of the Royal Society of London, B,* 334: 243–251.

Hemelrijk, C. K. 1996. Reciprocation in apes: From complex cognition to self-structuring. In *Great Ape Societies* (W. C. McGrew, L. F. Marchant, and T. Nishida, eds.), pp. 185–195. Cambridge: Cambridge University Press.

Herman, L. M., Pack, A. A., and Morrel-Samuels, P. 1993. Representational and conceptual skills of dolphins. In *Language and Communication: A Comparative Perspective* (H. L. Roitblat, L. M. Herman, and P. E. Nachtigall, eds.), Hillsdale, N.J.: Lawrence Erlbaum.

Herzog, M., and Hopf, S. 1984. Behavioral responses to species-specific warning calls in infant squirrel monkeys reared in social isolation. *American Journal of Primatology,* 7: 99–106.

Hewlett, B. S., and Cavalli-Sforza, L. L. 1986. Cultural transmission among Aka pygmies. *American Anthropologist,* 88: 922–934.

Hill, K., and Hawkes, K. 1983. Neotropical hunting among the Achè of eastern Paraguay. In *Adaptive Responses of Native Amazonians* (R. B. Hames and W. T. Vickers, eds.), pp. 139–188. New York: Academic Press.

Hill, R. A., and Dunbar, R. I. M. 1998. An evaluation of the roles of predation rate and predation risk as selective pressures on primate grouping behaviour. *Behaviour,* 135: 411–430.

Hill, R. A., and Lee, P. C. 1998. Predation risk as an influence on group size in Cercopithecoid primates: Implications for social structure. *Journal of the Zoological Society of London,* 245: 447–456.

Hinde, R. A. 1970. Aggression. In *Biology and the Human Sciences* (J. Pringle, ed.), pp. 1–23. Oxford: Clarendon.

Hiraiwa-Hasegawa, M., and Hasegawa, T. 1994. Infanticide in nonhuman primates: Sexual selection and local resource competition. In *Infanticide and Parental Care* (S. Parmigiani and F. S. vom Saal, eds.), pp. 137–154. Langhorne, Pa.: Harwood Academy Publishers.

Hockett, C. F. 1963. The problem of universals in language. In *Universals of Language* (J. H. Greenberg, ed.), pp. 1–29. Cambridge, Mass.: MIT Press.

Hodos, W., and Campbell, C. B. 1969. *Scala naturae;* Why there is no theory in comparative psychology. *Psychological Review,* 76: 337–350.

Hohmann, G., and Fruth, B. 1993. Field observations on meat sharing among bonobos *(Pan paniscus). Folia Primatologica,* 60: 225–229.

van Hooff, J. A. R. A. M. 1972. A comparative approach to the phylogeny of laughter and smiling. In *Nonverbal Communication* (R. A. Hinde, ed.), pp. 12–53. Cambridge: Cambridge University Press.

van Hooff, J. A. R. A. M., and van Schaik, C. 1992. Cooperation in competition:

The ecology of primate bonds. In *Coalitions and Alliances in Humans and Other Animals* (H. A. Harcourt and F. B. M. de Waal, eds.), pp. 357–389. Oxford: Oxford University Press.

——— 1994. Male bonds: Affiliative relationships among nonhuman primate males. *Behaviour,* 130: 309–337.

Hopkins, W. D., Marino, L., Rilling, J., and MacGregor, L. 1998. A comparative study of asymmetries in the plenum temporale in primates as revealed by magnetic resonance imaging (MRI). *NeuroReport,* 9: 2913–18.

Hrdy, S. B. 1979. Infanticide among animals: A review, classification, and examination of the implications for the reproductive strategies of females. *Ethology and Sociobiology,* 1: 13–40.

——— 1981. *The Woman That Never Evolved.* Cambridge, Mass.: Harvard University Press.

Hrdy, S. B., Janson, C., and van Schaik, C. 1995. Infanticide: Let's not throw out the baby with the bath water. *Evolutionary Anthropology,* 3: 151–154.

Hrdy, S. B., and Whitten, P. L. 1987. Patterning of sexual activity. In *Primate Societies* (B. B. Smuts, D. L. Cheney, R. M. Seyfarth, T. T. Struhsaker, and R. W. Wrangham, eds.), pp. 370–384. Chicago: University of Chicago Press.

Humphrey, N. K. 1976. The social function of intellect. In *Growing Points in Ethology* (P. P. G. Bateson and R. A. Hinde, eds.), pp. 303–317. Cambridge: Cambridge University Press.

Huttenlocher, J., Haight, W., Bryk, A., Selzer, M., and Lyons, T. 1991. Early vocabulary growth: Relation to language input and gender. *Developmental Psychology,* 27: 236–248.

Isaac, G. L. 1978. The food-sharing behavior of proto-human hominids. *Scientific American,* 238: 90–108.

Itani, T., and Nishimura, A. 1973. The study of intrahuman culture in Japan. In *Precultural Primate Behavior* (E. W. Menzel, ed.), pp. 26–50. Basel: Karger.

Jakobson, R., and Halle, M. 1956. *Fundamentals of Language.* The Hague: Mouton.

Joffe, T. H., and Dunbar, R. I. M. 1997. Visual and socio-cognitive information processing in primate brain evolution. *Proceedings of the Royal Society, London, B,* 264: 1303–7.

Johnson, J. S., and Newport, E. L. 1989. Critical period effects in second language learning: The influence of maturational state on the acquisition of English as a second language. *Cognitive Psychology,* 21: 60–99.

Jolly, A. 1966. Lemur social behaviour and primate intelligence. *Science,* 153: 501–506.

Jones, S., Martin, R. D., and Pilbeam, D. 1992. *The Cambridge Encyclopaedia of Human Evolution.* Cambridge: Cambridge University Press.

Jouanjean-l'Antoene, A. 1997. Reciprocal interactions and the development of communication and language between parents and children. In *Social Influences on Vocal Development* (C. T. Snowdon and M. Hausberger, eds.), pp. 312–327. Cambridge: Cambridge University Press.

Joulian, F. 1996. Comparing chimpanzee and early hominid techniques: Some contributions to cultural and cognitive questions. In *Modelling the Human Mind* (P. Mellers and K. R. Gibson, eds.), pp. 173–199. Exeter: Short Run.

Judge, P. 1982. Redirection of aggression based on kinship in a captive group of pigtail macaques (abstract). *International Journal of Primatology,* 3: 301.

Judge, P. G., and de Waal, F. B. M. 1997. Rhesus monkey behaviour under diverse population densities: Coping with long-term crowding. *Animal Behaviour,* 54: 643–662.

Kanazawa, S. 1998. What facial part is important for Japanese monkeys *(Macaca fuscata)* in recognition of smiling and sad faces of humans (*Homo sapiens*)? *Journal of Comparative Psychology,* 112: 363–370.

Kano, T. 1992. *The Last Ape: Pygmy Chimpanzee Behavior and Ecology.* Stanford: Stanford University Press.

––––––– 1996. Male rank order and copulation rate in a unit-group of bonobos at Wamba, Zaire. In *Great Ape Societies* (W. C. McGrew, L. F. Marchant, and T. Nishida, eds.), pp. 135–145. Cambridge: Cambridge University Press.

––––––– 1998. Comments on C. B. Stanford. *Current Anthropology,* 39: 410–411.

Kawai, M., Watanabe, K., and Mori, A. 1992. Precultural behaviors observed in free-ranging Japanese monkeys on Koshima Islet over the past 25 years. *Primate Report,* 32: 143–153.

Kegl, J., Senghas, A., Coppola, M. 1996. Creation through contact: Sign language emergence and sign language change in Nicaragua. In *Comparative Grammatical Change: The Intersection of Language Acquisition, Creole Genesis and Diachronic Syntax* (M. deGraff, ed.), Cambridge, Mass.: MIT Press.

King, B. J. 1994. *The Information Continuum: Evolution of Social Information Transfer in Monkeys, Apes and Hominids.* Santa Fe, N. Mex.: School of American Research Press.

Kinzey, W. (ed.). 1987. *The Evolution of Human Behavior: Primate Models.* Albany: State University of New York Press.

Kluender, K. R. 1994. Speech perception as a tractable problem in cognitive science. In *Handbook of Psycholinguistics* (M. A. Gernsbacher, ed.), pp. 173–217. San Diego: Academic Press.

Kluender, K. R., Diehl, R. L., and Killeen, P. R. 1987. Japanese quail can learn phonetic categories. *Science,* 237: 1195–97.

Köhler, W. 1927. *The Mentality of Apes,* 2nd ed. London: Kegan Paul, Trench, Trubner.

Köhler, M., and Moyà-Solà, S. 1997. Ape-like or hominid-like? The positional behavior of *Oreopithecus bambolii* reconsidered. *Proceedings of the National Academy of Sciences USA,* 94: 11747–50.

Kortlandt, A. 1986. The use of stone tools by wild-living chimpanzees and earliest hominids. *Journal of Human Evolution,* 15: 77–132.

Kroeber, A. L. 1928. Sub-human cultural beginnings. *Quarterly Review of Biology,* 3: 325–342.

Kuhl, P. K., and Miller, J. D. 1975. Speech perception in the chinchilla: Voiced-voiceless distinction in alveolar plosive consonants. *Science,* 190: 69–72.

Kummer, H. 1967. Tripartite relations in hamadryas baboons. In *Social Communication among Primates* (S. A. Altmann, ed.), pp. 63–71. Chicago: University of Chicago Press.

———— 1982. Social knowledge in free-ranging primates. In *Animal Mind— Human Mind* (D. R. Griffin, ed.). New York: Springer-Verlag.

———— 1995. *In Quest of the Sacred Baboon*. Princeton: Princeton University Press.

Kummer, H., and Goodall, J. 1985. Conditions of innovative behaviour in primates. *Philosophical Transactions of the Royal Society of London, B,* 308: 203–214.

Kuroda, S. 1984. Interaction over food among pygmy chimpanzees. In *The Pygmy Chimpanzee* (R. L. Susman, ed.), pp. 301–324. New York: Plenum.

Leavens, D. A., and Hopkins, W. D. 1999. The whole-hand point: The structure and function of pointing from a comparative perspective. *Journal of Comparative Psychology*, 113: 417–425.

Lee, R. B., and DeVore, I. (eds.). 1968. *Man the Hunter*. Chicago: Aldine.

Liberman, A. M., Cooper, F. S., Shankweiler, D. P., and Studdert-Kennedy, M. 1967. Perception of the speech code. *Psychological Review*, 74: 431–461.

Lieberman, P. 1975. *On the Origins of Language: An Introduction to the Evolution of Human Speech*. New York: Macmillan.

Locke, J. 1993. *The Child's Path to Spoken Language*. Cambridge, Mass.: Harvard University Press.

Locke, J. L., and Snow, C. 1997. Social influences on vocal learning in human and nonhuman primates. In *Social Influences on Vocal Development* (C. T. Snowdon and M. Hausberger, eds.), pp. 274–292. Cambridge: Cambridge University Press.

Lorenz, K. Z. 1966 [1963]. *On Aggression*. London: Methuen.

Lovejoy, C. O. 1981. The origin of man. *Science*, 211: 341–350.

MacDonald, D. 1984. *The Encyclopedia of Mammals*. New York: Facts on File Publications.

Machiavelli, N. 1979 [1532]. *The Prince*. Harmondsworth, Middlesex: Penguin Books.

Macphail, E. 1982. *Brain and Intelligence in Vertebrates*. Oxford: Clarendon Press.

Maddieson, I. 1984. *Patterns of Sound*. Cambridge: Cambridge University Press.

Maestripieri, D. 1995. Maternal encouragement in nonhuman primates and the question of animal teaching. *Human Nature*, 6: 361–378.

Malenky, R. K., and Wrangham, R. W. 1994. A quantitative comparison of terrestrial herbaceous food consumption by *Pan paniscus* in the Lomako Forest, Zaire, and *Pan troglodytes* in the Kibale Forest, Uganda. *American Journal of Primatology*, 32: 1–12.

Malinowski, B. 1929. *The Sexual Life of Savages*. London: Lowe and Brydone.

Marks, J. 1994. Blood will tell (won't it?): A century of molecular discourse in anthropological systematics. *American Journal of Physical Anthropology*, 94: 59–79.

Marler, P. 1970. Birdsong and speech development: Could there be parallels? *American Scientist*, 58: 669–674.

Marler, P., and Peters, S. 1982. Subsong and plastic song: Their role in vocal learning processes. In *Acoustic Communication in Birds*, Vol. 1, *Song Learning and Its Development* (D. E. Kroodsma and E. H. Miller, eds.), pp. 25–50. New York: Academic Press.

Martin, R. D. 1992. Female cycles in relation to paternity in primate societies. In *Paternity in Primates: Genetic Tests and Theories* (R. D. Martin, A. F. Dixson, and E. J. Wickings, eds.), pp. 238–274. Basel: Karger.

Masataka, N. 1983. Categorical responses to natural and synthesized alarm calls in Goeldi's monkeys *(Callimico goeldi)*. *Primates*, 24: 40–51.

Matsuzawa, T. 1994. Field experiments on use of stone tools by chimpanzees in the wild. In *Chimpanzee Cultures* (R. W. Wrangham, F. B. M. de Waal, W. C. McGrew, and P. G. Heltne, eds.), pp. 351–370. Cambridge, Mass.: Harvard University Press.

———— 1996. Chimpanzee intelligence in nature and in captivity: Isomorphism of symbol use and tool use. In *Great Ape Societies* (W. C. McGrew, L. F. Marchant, and T. Nishida, eds.), pp. 196–209. Cambridge: Cambridge University Press.

May, B., Moody, D., and Stebbins, W. 1989. Categorical perception of conspecific communication sounds by Japanese macaques. *Journal of the Acoustical Society of America*, 85: 837–847.

McConnell, P. B. 1990. Acoustic structure and receiver response in *Canis familiaris*. *Animal Behaviour*, 39: 897–904.

McConnell, P. B., and Snowdon, C. T. 1986. Vocal interactions between unfamiliar groups of captive cotton top tamarins. *Behaviour*, 97: 273–296.

McCune, L., Vihman, M. M., Roug-Hellichius, L., Delery, D. B., and Gogate, L. 1996. Grunt communication in human infants *(Homo sapiens)*. *Journal of Comparative Psychology*, 110: 27–37.

McGrew, W. C. 1989. Why is ape tool use so confusing? In *Comparative Socioecology: The Behavioural Ecology of Humans and Other Mammals* (V. Standen and R. A. Foley, eds.), pp. 457–472. Oxford: Blackwell Scientific.

———— 1992. *Chimpanzee Material Culture: Implications for Human Evolution.* Cambridge: Cambridge University Press.

———— 1998. Culture in non-human primates? *Annual Review of Anthropology*, 27: 301–328.

McGrew, W. C., Ham, R. M., White, L. J. T., Tutin, C. E. G., and Fernandez, M. 1997. Why don't chimpanzees in Gabon crack nuts? *International Journal of Primatology*, 18: 353–374.

McGrew, W. C., and Marchant, L. F. 1997. Using the tools at hand: Manual laterality and elementary technology in *Cebus* spp. and *Pan* spp. *International Journal of Primatology*, 18: 787–810.

McGrew, W. C., Marchant, L. F., and Nishida, T. 1996. *Great Ape Societies.* Cambridge: Cambridge University Press.

Mellars, P., and Gibson, K. 1996. *Modelling the Early Human Mind.* Cambridge: McDonald Institute for Archaeological Research.

Menzel, E. W. (ed.) 1973. *Precultural Primate Behavior.* Basel: Karger.

Mesnick, S. 1997. Sexual alliances: Evidence and evolutionary implications. In *Feminism and Evolutionary Biology: Boundaries, Intersections, and Frontiers* (P. A. Gowaty, ed.), pp. 207–260. New York: Chapman and Hall.

Mitani, J. 1996. Comparative studies of African ape vocal behavior. In *Great Ape Societies* (W. C. McGrew, L. F. Marchant, and T. Nishida, eds.), pp. 241–254. Cambridge: Cambridge University Press.

Mitchell, C. L. 1994. Migration alliances and coalitions among adult male South American squirrel monkeys *(Saimiri sciureus)*. *Behaviour,* 130: 169–190.

Montagu, A. 1976. *The Nature of Human Aggression.* New York: Oxford University Press.

Moore, J. 1984. Female transfer in primates. *International Journal of Primatology,* 5: 537–589.

———— 1992. Dispersal, nepotism, and primate social behavior. *International Journal of Primatology,* 13: 361–378.

Moore, J., and Ali, R. 1984. Are dispersal and inbreeding avoidance related? *Animal Behaviour,* 32: 94–112.

Morin, P. A., Moore, J. J., et al. 1994. Kin selection, social structure, gene flow, and the evolution of chimpanzees. *Science,* 265: 1193–1201.

Morris, D. 1967. *The Naked Ape.* New York: Dell.

Morse, P. A., and Snowdon, C. T. 1975. An investigation of categorical speech discrimination by rhesus monkeys. *Perception and Psychophysics,* 17: 9–16.

Mundinger, P. 1970. Vocal imitation and recognition of finch calls. *Science,* 168: 480–482.

Nakamura, M., McGrew, W. C., and Marchant, L. F. 1999. Social scratch: Another custom in wild chimpanzees? *Primates* (in press).

Napier, J., and Napier, P. 1985. *The Natural History of the Primates.* Cambridge, Mass.: MIT Press.

Nelson, D. A., and Marler, P. 1989. Categorical perception of a natural stimulus continuum: Birdsong. *Science,* 244: 976–978.

Newell, A., Shaw, J. C., and Simon, H. A. 1958. Elements of a theory of human problem solving. *Psychological Review,* 65: 151–166.

Newman, J. D., and Symmes, D. 1982. Inheritance and experience in the acquisition of primate acoustic behavior. In *Primate Communication* (C. T. Snowdon, C. H. Brown, and M. R. Petersen, eds.), pp. 259–278. New York: Cambridge University Press.

Nimchinsky, E. A., Gilissen, E., Allman, J. M., Perl, D. P., Erwin, J. M., and Hof, P. R. 1999. A neuronal morphologic type unique to humans and great apes. *Proceedings of the National Academy of Sciences USA,* 96: 5268–73.

Nishida, T. 1968. The social group of wild chimpanzees in the Mahale Mountains. *Primates,* 9: 167–224.

———— 1979. The social structure of chimpanzees of the Mahale mountains. In *The Great Apes* (D. A. Hamburg and E. R. McCown, eds.), pp. 73–121. Menlo Park, Calif.: Benjamin/Cummings.

———— 1983. Alpha status and agonistic alliance in wild chimpanzees *(Pan troglodytes schweinfurthii)*. *Primates,* 24: 318–336.

——— 1987. Local traditions and cultural transmission. In *Primate Societies* (B. B. Smuts, D. L. Cheney, R. M. Seyfarth, R. W. Wrangham, and T. T. Struhsaker, eds.), pp. 462–474. Chicago: University of Chicago Press.

——— (ed.). 1990. *The Chimpanzees of the Mahale Mountains: Sexual and Life History Strategies.* Tokyo: University of Tokyo Press.

Nishida, T., and Kawanaka, S. 1985. Within-group cannibalism by adult male chimpanzees. *Primates,* 26: 274–284.

Noble, W., and Davidson, I. 1994. *Human Evolution, Language and the Mind: A Psychological and Archeological Inquiry.* Cambridge: Cambridge University Press.

Nowicki, S. 1989. Vocal plasticity in captive black-capped chickadees: The acoustic basis of call convergence. *Animal Behaviour,* 37: 64–73.

Oba, R., and Masataka, N. 1996. Interspecific responses of ringtailed lemurs to playback of antipredator alarm calls given by Verraux's sifakas. *Ethology,* 102: 441–453.

O'Connell, J. F., and Hawkes, K. 1988. Hadza hunting, butchering, and bone transport and their archaeological implications. *Journal of Anthropological Research,* 44: 113–161.

Omedes, A. 1985. Ontogeny of social communication in silvery marmosets, *Callithrix argentata. Miscellanea Zoologica,* 9: 413–418.

Owings, D. O., and Morton, E. S. 1998. *Animal Vocal Communication: A New Approach.* Cambridge: Cambridge University Press.

Owren, M. J., Seyfarth, R. M., and Hopp, S. L. 1992. Categorical vocal signaling in non-human primates. In *Nonverbal Vocal Communication: Comparative and Developmental Approaches* (H. Papousek, U. Jurgens, and M. Papousek, eds.), pp. 102–122. New York: Cambridge University Press.

Packer, C., Collins, D. A., et al. 1995. Reproductive constraints on aggressive competition in female baboons. *Nature,* 373: 60–63.

Palombit, R. A. 1994. Extra-pair copulations in a monogamous ape. *Animal Behavior,* 47: 721–723.

——— 1995. Longitudinal patterns of reproduction in wild female siamang *(Hylobates syndactylus)* and white-handed gibbons *(Hylobates lar). International Journal of Primatology,* 16: 739–760.

Parish, A. R. 1993. Sex and food control in the "uncommon chimpanzee": How bonobo females overcome a phylogenetic legacy of male dominance. *Ethology and Sociobiology,* 15: 157–179.

——— 1996. Female relationships in bonobos *(Pan paniscus):* Evidence for bonding, cooperation, and female dominance in a male-philopatric species. *Human Nature,* 7: 61–96.

Parish, A. R., and de Waal, F. B. M. 2000. The other "closest living relative": How bonobos *(Pan paniscus)* challenge traditional assumptions about females, dominance, intra- and inter-sexual interactions, and hominid evolution. In *Evolutionary Perspectives on Human Reproductive Behavior* (D. Le Croy and P. Moller, eds.) *Annals of the New York Academy of Sciences,* 907: 97–113.

———— In press. Social relationships in the bonobo *(Pan paniscus)* redefined: Evidence for female-bonding in a "non-female bonded" primate. *Behaviour.*

Parker, S. T. 1996. Apprenticeship in tool-mediated extractive foraging: The origins of imitation, teaching and self-awareness in great apes. In *Reaching into Thought* (A. Russon, K. Bard, and S. T. Parker, eds.), pp. 348–370. Cambridge: Cambridge University Press.

Parker, S. T., and Gibson, K. R. 1977. Object manipulation, tool use, and sensorimotor intelligence as feeding adaptations in cebus monkeys and great apes. *Journal of Human Evolution,* 6: 623–641.

Parmigiani, S., and vom Saal, F. S. (eds.) 1994. *Infanticide and Parental Care.* Chur: Harwood.

Parr, L. A., Dove, T., and Hopkins, W. D. 1998. Why faces may be special: Evidence of the inversion effect in chimpanzees. *Journal of Cognitive Neuroscience,* 10: 615–622.

Parr, L. A., Hopkins, W. D., and de Waal, F. B. M. 1999. The perception of facial expressions by chimpanzees, *Pan troglodytes. Evolution of Communication* (in press).

Passingham, R. E. 1981. Primate specializations in brain and intelligence. *Symposia of the Zoological Society of London,* 4: 361–388.

Pepperberg, I. M. 1997. Social influences on the acquisition of human-based codes in parrots and nonhuman primates. In *Social Influences on Vocal Development* (C. T. Snowdon and M. Hausberger, eds.), pp. 157–177. Cambridge: Cambridge University Press.

Petitto, L. A., and Martentette, P. F. 1991. Babbling in the manual mode: Evidence for the ontogeny of language. *Science,* 251: 1493–96.

Pilbeam, D. 1996. Genetic and morphological records of the Hominoidea and hominid origins: A synthesis. *Molecular Phylogenetics and Evolution,* 5: 155–168.

Pinker, S. 1994. *The Language Instinct.* New York: William Morrow.

Pola, Y. V., and Snowdon, C. T. 1975. The vocalizations of pygmy marmosets, *Cebuella pygmaea. Animal Behaviour,* 23: 823–846.

Pope, T. R. 1990. The reproductive consequences of cooperation in the red howler monkey: Paternity exclusion in multi-male and single-male troops using genetic markers. *Behavioral Ecology and Sociobiology,* 27: 439–446.

Potts, R. 1998. Environmental hypotheses of hominid evolution. *Yearbook of Physical Anthropology,* 41: 93–136.

Povinelli, D. J., and Cant, J. G. H. 1995. Arboreal clambering and the evolution of self-conception. *Quarterly Journal of Biology,* 70: 393–421.

Power, M. 1991. *The Egalitarians: Human and Chimpanzee.* Cambridge: Cambridge University Press.

Pusey, A. E. 1979. Intercommunity transfer of chimpanzees in Gombe National Park. In *The Great Apes* (D. A. Hamburg and E. R. McCown, eds.), pp. 465–479. Menlo Park, Calif.: Benjamin/Cummings.

———— 1980. Inbreeding avoidance in chimpanzees. *Animal Behaviour,* 28: 543.

——— 1987. Sex-biased dispersal and inbreeding avoidance in birds and mammals. *Trends in Ecology and Evolution*, 2: 295–299.

——— 1990a. Behavioural changes at adolescence in chimpanzees. *Behaviour*, 115: 203.

——— 1990b. Mechanisms of inbreeding avoidance in nonhuman primates. In *Pedophilia: Biosocial Dimensions* (J. R. Feierman, ed.), pp. 201–220. New York: Springer-Verlag.

Pusey, A. E., and Packer, C. 1987. Dispersal and philopatry. In *Primate Societies* (B. B. Smuts, D. L. Cheney, R. M. Seyfarth, T. T. Struhsaker, and R. W. Wrangham, eds.), pp. 250–266. Chicago: University of Chicago Press.

Pusey, A. E., Williams, J. M., et al. 1997. The influence of dominance rank on the reproductive success of female chimpanzees. *Science*, 277: 828–831.

Pusey, A. E., and Wolf, M. 1996. Inbreeding avoidance in animals. *Trends in Ecology and Evolution*, 11: 201–206.

Ralls, K., Ballou, J. D., and Templeton, A. 1988. Estimates of lethal equivalents and the cost of inbreeding in mammals. *Conservation Biology*, 2: 185–193.

Randall, J. A. 1995. Modification of foot-drumming signatures by kangaroo rats: Changing territories and gaining new neighbors. *Animal Behaviour*, 49: 1227–37.

Reichard, U., and Sommer, V. 1997. Group encounters in wild gibbons (*Hylobates lar*): Agonism, affiliation, and the concept of infanticide. *Behaviour*, 134: 1135–74.

Richard, A. F. 1981. Changing assumptions in primate ecology. *American Anthropologist*, 83: 517–533.

Robbins, M. M. 1995. A demographic analysis of male life history and social structure of mountain gorillas. *Behaviour*, 132: 21–48.

——— 1996. Male-male interactions in heterosexual and all-male wild mountain gorilla groups. *Ethology*, 102: 942–965.

Robinson, J. G. 1984. Syntactic structures in the vocalizations of wedge-capped capuchin monkeys, *Cebus olivaceus*. *Behaviour*, 46–79.

Rodseth, L., Wrangham, R. W., et al. 1991. The human community as a primate society. *Current Anthropology*, 32: 221–433.

Roosevelt, A. C. In press. Early human occupation in the equatorial rainforests: Congo Basin, Africa.

Ross, C. 1991. Life history patterns of New World monkeys. *International Journal of Primatology*, 12: 481–502.

Roush, R. S., and Snowdon, C. T. 1994. Ontogeny of food-associated calls in cotton-top tamarins. *Animal Behaviour*, 47: 263–273.

——— In press. The effects of social status on food-associated calling behaviour in captive cotton-top tamarins. *Animal Behaviour*.

Rowell, T. E. 1993. Reification of social systems. *Evolutionary Anthropology*, 2: 135–137.

Russon, A. E. 1997. Exploiting the expertise of others. In *Machiavellian Intelligence, II: Extensions and Evaluations* (A. Whiten and R. W. Byrne, eds.), pp. 174–206. Cambridge: Cambridge University Press.

Russon, A. E., Bard, K. A., and Parker, S. T. (eds.). 1996. *Reaching into Thought: The Minds of the Great Apes*. Cambridge: Cambridge University Press.

Russon, A. E., and Galdikas, B. 1995. Constraints on great apes' imitation: Model and action selectivity in rehabilitant orangutan *(Pongo pygmeus)* imitation. *Journal of Comparative Psychology*, 109: 5–17.

Saffran, J. R., Aslin, R. N., and Newport, E. L. 1996. Statistical learning in 8 month-old infants. *Science*, 274: 1926–28.

Sambrook, T., and Whiten, A. 1997. On the nature of complexity in cognitive and behavioural science. *Theory and Psychology*, 7: 191–213.

Savage-Rumbaugh, S., and Lewin, R. 1994. *Kanzi: The Ape at the Brink of the Human Mind*. New York: John Wiley.

Savage-Rumbaugh, S., Murphy, J., Sevchik, R. A., Brakke, K. E., Williams, S. A., and Rumbaugh, D. M. 1993. Language comprehension in ape and child. *Monographs of the Society for Research in Child Development*, 58: 1–254.

van Schaik, C. P., and Dunbar, R. I. M. 1990. The evolution of monogamy in large primates: A new hypothesis and some crucial tests. *Behaviour*, 115: 30–62.

van Schaik, C. P., Fox, E. A., and Sitompul, A. F. 1996. Manufacture and use of tools in wild Sumatran orangutans: Implications for human evolution. *Naturwissenschaften*, 83: 186–188.

Schick, K. D., and Toth, N. 1993. *Making Silent Stones Speak*. New York: Simon and Schuster.

Schwarz, E. 1929. Das Vorkommen des Schimpansen auf den linken Kongo-Ufer. *Revue de Zoologie et de Botanique Africaines*, 16: 425–426.

Seyfarth, R. M., and Cheney, D. 1984. Grooming alliances and reciprocal altruism in vervet monkeys. *Nature*, 308: 541–542.

—— 1997. Some general features of vocal development in nonhuman primates. In *Social Influences on Vocal Development* (C. T Snowdon and M. Hausberger, eds.), pp. 249–273. Cambridge: Cambridge University Press.

Shipman, P. 1986. Scavenging or hunting in early hominids. *American Anthropologist*, 88: 27–43.

Short, R. V. 1979. Sexual selection and its component parts, somatic and genital selection, as illustrated by man and the great apes. *Advances in the Study of Behavior*, 9.

Shreeve, J. 1996. Sunset on the savanna. *Discover*, 17: 116–125.

Small, M. F. 1988. Female primate sexual behavior and conception. *Current Anthropology*, 29: 81–100.

—— 1989. Female choice in nonhuman primates. *Yearbook of Physical Anthropology*, 32: 103–127.

—— 1992. Female choice in mating: The evolutionary significance of female choice depends on why the female chooses her reproductive partner. *American Scientist*, 80: 142–151.

Smuts, B. B. 1985. *Sex and Friendship in Baboons*. New York: Aldine.

—— 1995. The evolutionary origins of patriarchy. *Human Nature*, 6: 1–32.

Smuts, B. B., and Smuts, R. W. 1993. Male aggression and sexual coercion of females in nonhuman primates and other mammals: Evidence and theoretical implications. *Advances in the Study of Behaviour*, 22: 1–63.

Snowdon, C. T. 1987. A naturalistic view of categorical perception. In *Categorical Perception* (S. Harnad, ed.), pp. 332–354. New York: Cambridge University Press.

——— 1993. A vocal taxonomy of the Callitrichids. In *Marmosets and Tamarins: Systematics, Behaviour and Ecology* (A. B. Rylands, ed.), pp. 78–94. Oxford: Oxford University Press.

Snowdon, C. T., and Cleveland, J. 1984. "Conversations" among pygmy marmosets. *American Journal of Primatology*, 7: 15–20.

Snowdon, C. T., and Elowson, A. M. 1999. Pygmy marmosets modify call structure when paired. *Ethology* (in press).

Snowdon, C. T., Elowson, A. M., and Roush, R. S. 1997. Social influences on vocal development in New World primates. In *Social Influences on Vocal Development* (C. T. Snowdon and M. Hausberger, eds.), pp. 234–248. Cambridge: Cambridge University Press.

Snowdon, C. T., and Hausberger, M. (eds.) 1997. *Social Influences on Vocal Development*. Cambridge: Cambridge University Press.

Snowdon, C. T., and Hodun, A. 1981. Acoustic adaptations in pygmy marmoset contact calls: Locational cues vary with distance between conspecifics. *Behavioral Ecology and Sociobiology*, 9: 295–300.

Snowdon, C. T., and Pola, Y. V. 1978. Interspecific and intraspecific responses to synthesized marmoset vocalizations. *Animal Behaviour*, 26: 192–206.

Sorensen, A. P., Jr. 1967. Multilingualism in the Northwest Amazon. *American Anthropologist*, 69: 670–685.

Stanford, C. B. 1996. The hunting ecology of wild chimpanzees: Implications for the behavioral ecology of Pliocene hominids. *American Anthropologist*, 98: 96–113.

——— 1998a. *Chimpanzee and Red Colobus: The Ecology of Predator and Prey*. Cambridge, Mass.: Harvard University Press.

——— 1998b. The social behavior of chimpanzees and bonobos: Empirical evidence and shifting assumptions. *Current Anthropology*, 39: 399–420.

——— 1999. *The Hunting Apes: Meat Eating and the Origins of Human Behavior*. Princeton: Princeton University Press.

——— Forthcoming. Hunting apes: A comparison of social meat-foraging by chimpanzees and human foragers. In *Meat-Eating and Human Evolution* (C. B. Stanford and H. T. Bunn, eds.). Oxford: Oxford University Press.

Stanford, C. B., Wallis, J., Matama, H., and Goodall, J. 1994. Patterns of predation by chimpanzees on red colobus monkeys in Gombe National Park, Tanzania, 1982–1991. *American Journal of Physical Anthropology*, 94: 213–228.

Stanger, K. F. 1993. Structure and function of the vocalizations of nocturnal prosimians (Cheirogaleidae). Unpublished doctoral dissertation, Eberhard-Karls-Universität, Tübingen, Germany.

Strier, K. B. 1990. New World primates, new frontiers: Insights from the woolly spider monkey, or muriqui *(Brachyteles arachnoides). International Journal of Primatology,* 11: 7–19.

———— 1994. Myth of the typical primate. *Yearbook of Physical Anthropology,* 37: 233–271.

———— 1996. Male reproductive strategies in New World primates. *Human Nature,* 7: 105–123.

———— 1999a. Why is female kin bonding so rare: Comparative sociality of New World primates. In *Primate Socioecology* (P. Lee, ed.). Cambridge: Cambridge University Press.

———— 1999b. *Faces in the Forest: The Endangered Muriqui Monkeys of Brazil.* Cambridge, Mass.: Harvard University Press.

Strier, K. B., and Ziegler, T. E. 1997. Behavioral and endocrine characteristics of the reproductive cycle in wild muriqui monkeys, *Brachyteles arachnoides. American Journal of Primatology,* 42: 299–310.

Sugiyama, Y. 1984. Population dynamics of wild chimpanzees at Bossou, Guinea, between 1976 and 1983. *Primates,* 25: 391–400.

———— 1997. Social tradition and the use of tool-composites by wild chimpanzees. *Evolutionary Anthropology,* 6: 23–27.

Sugiyama, Y., Kawamoto, S., et al. 1993. Paternity discrimination and intergroup relationships of chimpanzees at Bossou. *Primates,* 34: 545–552.

Sugiyama, Y., and Koman, J. 1979. Social structure and dynamics of wild chimpanzees at Bossou, Guinea. *Primates,* 20: 323–329.

Susman, R. L. 1984a. The locomotor behavior of *Pan paniscus* in the Lomako Forest. In *The Pygmy Chimpanzee: Evolutionary Biology and Behavior* (R. L. Susman, ed.), pp. 369–393. New York: Plenum.

———— (ed.). 1984b. *The Pygmy Chimpanzee: Evolutionary Biology and Behavior.* New York: Plenum.

Sussman, R. W. 1992. Male life history and intergroup mobility among ringtailed lemurs *(Lemur catta). International Journal of Primatology,* 13: 395–413.

Sussman, R. W., Cheverud, J. M., and Bartlett, T. Q. 1995. Infant killing as an evolutionary strategy: Reality or myth? *Evolutionary Anthropology,* 3: 149–151.

Suzuki, S., Hill, D. A., Maruhashi, T., and Tsukahara, T. 1990. Frog- and lizard-eating behavior of wild Japanese macaques in Yakushima, Japan. *Primates,* 31: 421–426.

Taberlet, P., Wiats, L. P., and Luikart, G. 1999. Noninvasive genetic sampling: Look before you leap. *Trends in Ecology and Evolution,* 14: 323–327.

Takahata, Y., Ihobe, H., and Idani, G. 1996. Comparing copulations of chimpanzees and bonobos: Do females exhibit proceptivity or receptivity? In *Great Ape Societies* (W. C. McGrew, L. F. Marchant, and T. Nishida, eds.), pp. 146–155. Cambridge: Cambridge University Press.

Talmage-Riggs, G., Winter, P., Ploog, D., and Mayer, W. 1972. Effect of deafening on the vocal behavior of the squirrel monkey *(Saimiri sciureus)*. *Folia Primatologica*, 17: 404–420.

Tanaka, I. 1995. Matrilineal distribution of louse egg-handling techniques during grooming in free-ranging Japanese macaques. *American Journal of Physical Anthropology*, 98: 197–201.

Tenaza, R. 1985. Songs of hybrid gibbons *(Hylobates lar X H. muelleri)*. *American Journal of Primatology*, 8: 249–253.

Terkel, J. 1996. Cultural transmission of feeding behavior in the black rat *(Rattus rattus)*. In *Social Learning in Animals: The Roots of Culture* (C. M. Hayes, and B. G. Galef, eds.), pp. 17–47. San Diego: Academic Press.

Thompson-Handler, N. 1990. The pygmy chimpanzee: Sociosexual behavior, reproductive biology and life history patterns. Unpublished doctoral dissertation, Yale University, New Haven.

Tomasello, M., and Call, J. 1997. *Primate Cognition*. Oxford: Oxford University Press.

Tooby, J., and DeVore, I. 1987. The reconstruction of hominid behavioral evolution through strategic modelling. In *The Evolution of Human Behavior: Primate Models* (W. Kinzey, ed.), pp. 183–238. Albany: State University of New York Press.

Toth, N., and Schick, K. 1993. Early stone inductries and inferences regarding language and cognition. In *Tools, Language, and Cognition in Human Evolution* (K. R. Gibson and T. Ingold, eds.), pp. 346–362. Cambridge: Cambridge University Press.

Tratz, E. P., and Heck, H. 1954. Der afrikanische Anthropoide "Bonobo," eine neue Menschenaffengattung. *Säugetierkundliche Mitteilungen*, 2: 97–101.

Trivers, R. L. 1985. *Social Evolution*. Menlo Park, Calif.: Benjamin/Cummings.

Tutin, C. E. G., and Fernandez, M. 1992. Insect-eating by sympatric lowland gorillas *(Gorilla g. gorilla)* and chimpanzees *(Pan t. troglodytes)* in the Lopè Reserve, Gabon. *American Journal of Primatology*, 28: 29–40.

Tutin, C. E. G., and McGinnis, P. R. 1981. Chimpanzee reproduction in the wild. In *Reproductive Biology of the Great Apes* (C. E. Graham, ed.). New York: Academic Press.

Tyack, P. L., and Sayigh, L. S. 1997. Vocal learning in cetaceans. In *Social Influences on Vocal Development* (C. T. Snowdon and M. Hausberger, eds.), pp. 208–233. Cambridge: Cambridge University Press.

Tylor, E. B. 1871. *Primitive Culture*. London: Murray.

Uehara, S. 1997. Predation on mammals by the chimpanzee *(Pan troglodytes)*: An ecological review. *Primates*, 38: 193–213.

Visalberghi, E. 1997. Success and understanding in cognitive tasks: A comparison between *Cebus apella* and *Pan troglodytes*. *International Journal of Primatology*, 18: 811–830.

Volman, S. F., and Khanna, H. 1995. Convergence of untutored song in group-reared zebra finches. *Journal of Comparative Psychology*, 109: 211–221.

de Waal, F. B. M. 1987. Tension regulation and nonreproductive functions of sex among captive bonobos *(Pan paniscus)*. *National Geographic Research*, 3: 318–335.

——— 1988. The communicative repertoire of captive bonobos *(Pan paniscus)*, compared to that of chimpanzees. *Behaviour*, 106: 183–251.

——— 1989. *Peacemaking among Primates*. Cambridge, Mass.: Harvard University Press.

——— 1991. Rank distance as a central feature of rhesus monkey social organisation: A sociometric analysis. *Animal Behaviour*, 41: 383–395.

——— 1992a. Aggression as a well-integrated part of primate social relationships: Critical comments on the Seville Statement on Violence. In *Aggression and Peacefulness in Humans and Other Primates* (J. Silverberg and J. P. Gray, eds.), pp. 37–56. New York: Oxford University Press.

——— 1992b. Appeasement, celebration, and food sharing in the two *Pan* species. In *Topics in Primatology*, Vol. 1, *Human Origins* (T. Nishida, W. C. McGrew, P. Marler, M. Pickford, and F. B. M. de Waal, eds.), pp. 37–50. Tokyo: University of Tokyo Press.

——— 1994. The chimpanzee's adaptive potential: A comparison of social life under captive and wild conditions. In *Chimpanzee Cultures* (R. W. Wrangham, W. C. McGrew, F. B. M. de Waal, and P. Heltne, eds.), pp. 243–260. Cambridge, Mass.: Harvard University Press.

——— 1995. Sex as an alternative to aggression in the bonobo. In *Sexual Nature, Sexual Culture* (P. Abramson and S. Pinkerton, eds.), pp. 37–56. Chicago: University of Chicago Press.

——— 1996. *Good Natured: The Origins of Right and Wrong in Humans and Other Animals*. Cambridge, Mass.: Harvard University Press.

——— 1997a. *Bonobo: The Forgotten Ape*. Berkeley: University of California Press.

——— 1997b. The chimpanzee's service economy: Food for grooming. *Evolution and Human Behavior*, 18: 375–386.

——— 1998 [1982]. *Chimpanzee Politics: Power and Sex among Apes*. Baltimore: Johns Hopkins University Press.

de Waal, F. B. M., and Aureli, F. 1996. Consolation, reconciliation, and a possible cognitive difference between macaque and chimpanzee. In *Reaching into Thought: The Minds of the Great Apes* (A. E. Russon, K. A. Bard, and S. T. Parker, eds.), pp. 80–110. Cambridge: Cambridge University Press.

de Waal, F. B. M., and van Roosmalen, A. 1979. Reconciliation and consolation among chimpanzees. *Behavioral Ecology and Sociobiology*, 5, 55–56.

Walker, A., and Shipman, P. 1996. *The Wisdom of the Bones*. New York: Knopf.

Wallis, J. 1997. A survey of reproductive parameters in the free-ranging chimpanzees of Gombe National Park. *Journal of Reproduction and Fertility*, 109: 297–307.

Wallis, J., and Goodall, J. 1993. Genital swelling patterns of pregnant chimpanzees in Gombe National Park. *American Journal of Primatology*, 10: 171–183.

Warren, J. M. 1973. Learning in vertebrates. In *Comparative Psychology: A Modern*

Survey (D. A. Dewsbury and D. A. Rethlingshafer, eds.), pp. 471–509. New York: McGraw Hill.

Washburn, S. L. 1978. What we can't learn about people from apes. *Human Nature*, 1(11): 70–75.

Watanabe, K. 1994. Precultural behavior of Japanese macaques: Longitudinal studies of the Koshima troops. In *The Ethological Roots of Culture* (R. A. Gardner, B. T. Gardner, B. Chiarelli, and F. X. Plooij, eds.), pp. 81–94. Dordrecht: Kluwer.

Waters, R. S., and Wilson, W. A., Jr. 1976. Speech perception by rhesus monkeys: The voicing distinction in synthesized labial and velar stop consonants. *Perception and Psychophysics*, 19: 285–289.

Watts, D. P. 1996. Comparative socio-ecology of gorillas. In *Great Ape Societies* (W. C. McGrew, L. F. Marchant, and T. Nishida, eds.), pp. 16–28. Cambridge: Cambridge University Press.

———— 1998. Coalitionary mate guarding by male chimpanzees at Ngogo, Kibale National Park. *Behavioral Ecology and Sociobiology*, 44: 43–55.

West, M. J., and King, A. P. 1996. Social learning: Synergy and songbirds. In *Social Learning in Animals: The Roots of Culture* (C. M. Heyes, and B. G. Galef, eds.), pp. 155–178. San Diego: Academic Press.

Westergaard, G. C. 1998. What capuchin monkeys can tell us about the origins of hominid material culture. *Journal of Material Culture*, 3: 5–19.

White, F. J., and Wrangham, R. W. 1988. Feeding competition and patch size in the chimpanzee species *Pan paniscus* and *P. troglodytes*. *Behaviour*, 105: 148–164.

Whiten, A. 1998. Imitation of the sequential structure of actions in chimpanzees *(Pan troglodytes)*. *Journal of Comparative Psychology*, 112: 270–281.

Whiten, A., and Byrne, R. W. 1988. Tactical deception in primates. *Behavioural and Brain Sciences*, 11: 233–273.

———— (eds.) 1997. *Machiavellian Intelligence, II: Extensions and Evaluations*. Cambridge: Cambridge University Press.

Whiten, A., Goodall, J., McGrew, W. C., Nishida, T., Reynolds, V., Sugiyama, Y., Tutin, C. E. G., Wrangham, R. W., and Boesch, C. (1999). Cultures in chimpanzees. *Nature*, 399: 682–685.

Whiten, A., and Ham, R. M. 1992. On the nature and evolution of imitation in the animal kingdom: Reappraisal of a century of research. *Advances in the Study of Behavior*, 21: 239–283.

Williamson, D., and Dunbar, R. I. M. 1999. Energetics, time budgets and group size. In *Comparative Primate Socioecology* (P. C. Lee, ed.). Cambridge: Cambridge University Press.

Wilson, E. O. 1978. *On Human Nature*. Cambridge, Mass.: Harvard University Press.

Wolpoff, M. 1998. *Paleoanthropology*. New York: Knopf.

Wood, B., and Collard, M. 1999. The human genus. *Science*, 284: 65–71.

Wrangham, R. W. 1979a. On the evolution of ape social systems. *Social Science Information*, 18(3): 335–368.

———— 1979b. Sex differences in chimpanzee dispersion. In *The Great Apes* (D. A. Hamburg and E. R. McCown, eds.), pp. 504–537. Menlo Park, Calif.: Benjamin/Cummings.

———— 1987. The significance of African apes for reconstructing human social evolution. In *The Evolution of Human Behavior: Primate Models* (W. G. Kinzey, ed.), pp. 51–71. Albany: State University of New York Press.

———— 1993. The evolution of sexuality in chimpanzees and bonobos. *Human Nature,* 4: 47–79.

———— 1997. Subtle, secret chimpanzees. *Science,* 277: 774–775.

———— 1999. The evolution of coalitionary aggression. *Yearbook of Physical Anthropology* (forthcoming).

Wrangham, R. W., and Clark, A. P., et al. 1992. Female social relationships and social organization of Kibale Forest chimps. In *Human Origins* (T. Nishida, W. C. McGrew, P. Marler, M. Pickford, and F. de Waal, eds.), pp. 81–98. Tokyo: University of Tokyo Press.

Wrangham, R. W., Jones, J., Laden, G., Pilbeam, D., and Conklin-Brittain, N. L. 1999. The raw and the stolen: Cooking and the ecology of human origins. *Current Anthropology* (in press).

Wrangham, R. W., McGrew, W. C., de Waal, F. B. M., and Heltne, P. G. (eds.) 1994. *Chimpanzee Cultures.* Cambridge, Mass.: Harvard University Press, with the Chicago Academy of Sciences.

Wrangham, R. W., and Peterson, D. 1996. *Demonic Males: Apes and the Evolution of Human Aggression.* Boston: Houghton Mifflin.

Yerkes, R. M. 1941. Conjugal contrasts among chimpanzees. *Journal of Abnormal Social Psychology,* 36: 175–199.

Zeh, J. A., and Zeh, D. W. 1996. The evolution of polyandry. I: Intragenomic conflict and genetic incompatibility. *Proceedings of the Royal Society, London, Series B.,* 263: 1711–17.

———— 1997. The evolution of polyandry. II: Post-copulatory defences against genetic incompatibility. *Proceedings of the Royal Society, London, Series B.,* 264: 69–75.

Ziegler, T. E., Santos, C. V., Pissinatti, A., and Strier, K. B. 1997. Steroid excretion during the ovarian cycle in captive and wild muriquis, *Brachyteles arachnoides. American Journal of Primatology,* 42: 311–321.

Zihlman, A. L. 1984. Body build and tissue composition in *Pan paniscus* and *Pan troglodytes,* with comparisons to other hominoids. In *The Pygmy Chimpanzee* (R. L. Susman, ed.), pp. 179–200. New York: Plenum.

Zihlman, A. L., Cronin, J. E., Cramer, D. L., and Sarich, V. M. 1978. Pygmy chimpanzee as a possible prototype for the common ancestor of humans, chimpanzees, and gorillas. *Nature,* 275: 744–746.

Zuckerman, S. 1932. *The Social Life of Monkeys and Apes.* London: Kegan, Paul, Trench, Trubner.

Contributors

RICHARD W. BYRNE
School of Psychology, University of Saint Andrews, Saint Andrews, United Kingdom

ROBIN I. M. DUNBAR
School of Biological Sciences, University of Liverpool, Liverpool, United Kingdom

WILLIAM C. MCGREW
Department of Sociology, Gerontology, and Anthropology, Miami University, Oxford, Ohio

ANNE E. PUSEY
Department of Evolution, Ecology, and Behavior, University of Minnesota, Saint Paul, Minnesota

CHARLES T. SNOWDON
Department of Psychology, University of Wisconsin, Madison, Wisconsin

CRAIG B. STANFORD
Department of Anthropology, University of Southern California, Los Angeles, California

KAREN B. STRIER
Department of Anthropology, University of Wisconsin, Madison, Wisconsin

FRANS B. M. DE WAAL
Living Links, Yerkes Primate Center, and Department of Psychology, Emory University, Atlanta, Georgia

RICHARD W. WRANGHAM
Department of Anthropology, Harvard University, Cambridge, Massachusetts

CPSIA information can be obtained
at www.ICGtesting.com
Printed in the USA
JSHW031034080721
16706JS00001B/21

9 780674 010048